DEVELOPMENT IN DISASTER-PRONE PLACES

For
Sarah, Brett and Kim

Development in disaster-prone places

Studies of vulnerability

JAMES LEWIS

INTERMEDIATE TECHNOLOGY PUBLICATIONS 1999

Intermediate Technology Publications Ltd
103–105 Southampton Row, London WC1B 4HH, UK

A CIP record for this publication is available from the British Library

ISBN 1 85339 472 6

Typeset by J&L Composition Ltd, Filey, North Yorkshire
Printed in the UK by The Cromwell Press, Trowbridge

Contents

Foreword

The division between crisis and disaster response on one hand, and development on the other, has long marred the efficiency of efforts to prevent damage in advance of disasters or to strengthen the impact of post-disaster actions, so that these contribute to reduced vulnerability in the future. Disasters occur in a long-term and local context, and it is unrealistic to assume a separation between 'normal' existence and those – often frequently recurring – periods that are disasters. Vulnerability, as this book shows, is an ongoing state. It needs to be addressed as such.

The over-narrow vision that many agencies and authorities charged with disaster preparedness and response have too often placed on the imminence and immediate aftermath of a disaster is a mistake, despite the argument that the means available and the urgency justify this limitation of scope. This narrow vision overlooks, or at least pays insufficient attention to, the broad-based and evolutive nature of vulnerability in its local context, and does not explore fully the circumstances in which disasters occur. Vulnerability can be addressed only through a co-ordinated and integrated strategy of preparation and response that considers the entire local context and its development, over time.

James Lewis's book makes an important contribution to these issues. First, it states clearly that the development process and the relief process must not be separated. Secondly, it brings an important and long under-valued argument to the fore, i.e. that 'a disaster is not a physical happening, it is a social event' (Quarantelli, 1986). This being the case, actions to reduce vulnerability and the subsequent impact of disaster must embrace social, economic, and political contexts, as well as the material and technical concerns and 'solutions' which tend to dominate, often because they are easier to quantify and handle.

The book illustrates through careful and thorough argument and telling case studies that vulnerability to disaster requires a locally developed strategy based on recognition of the full range of factors that contribute to vulnerability. Most interestingly, James Lewis draws attention to the intimate quality of vulnerability and disaster, the fact that disasters, when they strike, are a personal, family, community event, and only then a national or regional event. The suffering is individual and local. In the final resort, the resources – human and material – to reduce vulnerability, must also be local. This is a poignant expression of the reality that those disasters that make the news are not necessarily the most significant. The implication is clearly that one cannot rely on international response to small-scale events of potentially immense local importance.

For many years James Lewis has seemed at times to be a lone voice urging us to consider the broad vision of vulnerability, while others have focused on the more tangible area of disaster response. It is time for his message to be heard.

John Norton, Development Workshop, France June 1999

Preface

This book is one person's contribution to an internationally active and multi-disciplinary subject. The writer is also a researcher, 'individual consultant' and sole practitioner – albeit from time to time a 'team leader'. He is necessarily a generalist – and a generalist is a specialist in his own right. As more and more specialists are spawned, there is greater need for a generalist or two.

Academic roles usually allow or necessitate personal selection of subject area and its specialist pursuit. Consultancy, on the other hand, must largely and necessarily follow a path set by the inquiries, objectives or policies of others – which come to be modified (if at all) after the contribution of consultants. I have never been asked to undertake a task about which I had misgivings or which I felt I could not satisfactorily undertake or which I did not want to do. This says a lot for those who identify consultants – although some may have had their misgivings afterwards! One task, however, has often fortuitously led to another in a most satisfying way. Each subsequent consultancy has drawn upon previous work – each client has unknowingly contributed to the purposes of the next!

Writing for journal publication as a by-product of most assignments undertaken – insofar as contractual obligations allowed – resulted in work which is essentially 'experiential'. Research developed as Leverhulme Senior Research Fellow and co-founder and leader of the Disaster Research Unit at the University of Bradford, and similarly in the Centre for Development Studies at the University of Bath, where participation continues as Visiting Fellow in Development Studies.

Nevertheless, the series of studies forming the central part of this book, although expressing a continuation and progression, are not the product of an established academic programme, nor are they the by-product of an academic occupation.

From these multifarious sources there is a logical progression. The book places the studies together and in an ordered sequence, having introduced their overall subject of vulnerability in Part 1, and draws conclusions from them and other work in Part 3.

Influences upon this continuing work have been many and varied and, as custom has it, 'too many to mention by name'. Those who should not escape are Phil O'Keefe, Ken Westgate and Sue Jeffery who, already by then established in their respective disciplines, tolerated me at Bradford and Bath; Geof Wood and colleagues likewise at the Institute for International Policy Analysis, University of Bath (for more years than they care to remember), and the writings of the Natural Hazards Group at the Universities of Colorado and Toronto, references to which appear in the text and which it would be inappropriate to omit here because they indicate something of the long history of the study of 'natural hazards'.

Very belatedly, my gratitude also goes to the Leverhulme Trust, which funded my work at the Universities of Bradford and Bath. This book is a product of those times, as well as of work undertaken since – which would not have been possible were it not for work facilitated by the Trust. Though I am conscious of the time that it has taken to publish more than an occasional paper or two, I otherwise make no apology for references to work of 15 or more years ago – most of the processes and conditions of vulnerability to which this book refers are regrettably as extant now as they were then.

In addition to all of these have been the encounters with the many and various places, situations and experiences, and the many hundreds of people in the meeting rooms, houses and fields, villages and cities, institutions and governments, and on the aeroplanes, landrovers, carts, cars, trains, buses, bicycles and boats, with whom it has been my boundless privilege and pleasure to be all too briefly associated.

My wife Sarah, who follows her own career in which she is also an author, is another who cannot be permitted to escape the briefest mention. It is she and a young family who tolerated and coped with my absences, always, as it turned out, at crucial times. It is she to whom I turn initially with a new idea or to test the sense of something written – and it is she from whom many ideas and thoughts have originated. Above all, it is she who is the crucial source of encouragement when the point of what is being done has sometimes seemed so remote and difficult.

The onus is upon me for what I have done or not done in the light of these many opportunities and influences; I can only hope that I have given them deserved justice and that the small outcome presented here may be of some continued interest and usefulness.

Acknowledgements

I wish to thank the following for the permission to use the quoted material: John Murray (Publishers) Ltd for 'A Gilbertese Song' from *Return to the Islands*, translated by Arthur Grimble, on page viii; Taylor and Francis for the extract from *The Idea of Calamity* by Kenneth Hewitt on page xiii; Farrar, Strauss and Giroux, Book Publishers and Vintage Books for the extract from *A Small Place* by Jamaica Kincaid, on page 43; and Professor Roger McLean for the extract from *Island Reports No 1*, on page 67.

Introduction

THE RELATIONSHIP BETWEEN natural disasters and development is a recurring issue. On the one hand, development is regarded as deterministic and a major cause of vulnerability and the disasters it exacerbates; on the other hand, it is regarded (so far with less articulation) as the necessarily inevitable and appropriate vehicle for vulnerability and disaster reduction.

That vulnerability is the root cause of disasters is no longer a new issue nor is it a contentious one. This book seeks to explore and to expose some of the processes that have led to conditions of vulnerability. In doing so, it becomes evident that disasters are a contextual matter and that prevailing institutional, social and economic, as well as physical conditions, will largely determine their severity for populations and communities.

Vulnerability accrues as a result of processes of change and therefore is a potential product of all activities and undertakings of society. Vulnerability reduction therefore requires a multisectoral and pervasive responsibility that can comprehend and identify those activities and undertakings and implement their redirection. Such modifications to processes of change are a necessary component of development.

This means that disasters cannot be regarded as discrete events, because by doing so they become externalized from the activities and processes that create their context. Disasters are more usefully regarded as extensions of a pervasive normal hazardousness, because normal hazardousness is a comprehensible part of normal contexts. If we are going to be able to do anything at all about abnormal hazardousness, we should be attending to normal hazardousness, and our vulnerability to it, which is a part of everyday normality.

Development has often been identified as one of the contributors to processes of vulnerability; it is so again in this book. However, instead of all development being derogatorily regarded as a force for environmental determinism, the reverse of the same token is that development initiatives can be moulded to become the most appropriate means for vulnerability reduction.

Whether or not natural disasters are discrete events or 'extensions of a normal hazardousness' was debated at the commencement of natural hazards research in the USA (Burton *et al.*, 1968). *Natural hazards* distinguished that programme from the work of *natural disasters* research, which came later. Much of what followed in natural disasters research ignored or bypassed those issues. It was inherent in natural hazards research that there should be exploration of the interface between hazards and contexts; it was inherent in much of disasters response, if not in research, that contexts were self-evident; disasters were quite obviously different and separate and were therefore discrete events!

Natural hazards are those phenomena in nature that have the potential for causing damage; where that damage is caused to human populations and their settlements, disasters ensue. The title of this book might more correctly be

Development in hazard-prone places because it is the objective of the book to demonstrate that good development could reduce the risk of disasters in those places. In reality, however, disasters will have previously occurred in the hazard-prone places to which this book is addressed.

The book is based upon the viewpoint that there are disasters, but they are part of their contexts; they may be major but they are not discrete. Not only are they a part of their contexts, they are in part caused by their contexts.

Contexts are not only physical but social, institutional and political as well. That disasters may be caused by their contexts, causes the institutions of those contexts to react. The very existence, purpose and formation of institutions can be questioned (Hewitt, 1983) but how these institutions react reflects further on how the problem is managed. There is a useful comparison here with mental illness, another phenomenon which society today would rather did not exist. Until the seventeenth century, madmen lived alongside their sane fellows; madness was an ever-present force which was not excluded from everyday life (Foucault, quoted in Skultans, 1979). To what degree mental illness is a product of its context is not the point at issue here; but it is nevertheless useful to compare how madness came first to be invented and then treated, in descriptions of 'the dominant view' of hazards (in Hewitt, 1983):

> Natural calamity in a technocratic society is much the same sort of pivotal dilemma as insanity for the champions of reason. Disaster in the 20th-century international system involves comparable pressures upon dominant institutions and knowledge, as did the 'crazed poor' in the social and economic crises that formed the underbelly of the Enlightenment. Madness and calamity are very disturbing. They directly challenge our notions of order. . . They can be seen as clear limits to knowledge and power, because they are initiated in a way that seems uncontrollable by society.

> In both cases, however, there has arisen a dominant view that counteracts these difficulties with a positive creed, an assertion of potency where the grounds for conviction seem the least. It is exactly here that we can see the benefits to a technocratic approach of dividing off hazards. It is very convenient to treat calamity as a special problem. . . The problem is made manageable by an extreme narrowing of the range of interpretation and acceptable evidence. The resulting partial view has been achieved much as the 'great confinement' of the poverty stricken and 'crazy' in the 18th century was to form the foundation for dominant perspectives on madness, crime and punishment. . . We are not only dealing with the substance of such questions, or with a particular philosophy and set of practical procedures. We are also dealing with a careful, pragmatic and disarming *placement* of the problem. . .

> What emerges is that 'hazards ' are not viewed as integral parts of the spectrum of man–environment relations or as directly dependent upon those.

Nowadays, we know much more about what was called 'madness'; we know that it comprises a variety of identifiable conditions to which we allocate names such as depression, psychosis, schizophrenia. We also know that shades of these and other conditions exist in a normality that we all share in one way or another, but which are not recognized, not identified and not given a name or treatment.

These are part of our everyday usual condition. For some, the condition becomes abnormal, is diagnosed and given a name which then separates the condition (and sometimes the patient) from the normal. The experience may be nothing short of disaster, but people at large can then dissociate themselves from that unusualness. 'Emergency assistance' may have been required, and patients can and do 'rehabilitate' sufficiently to 'reconstruct' their lives.

This book is concerned with those experiences of 'disaster' which represent an intermediate transition between 'normal hazardousness' and major disaster. If vulnerability to small and localized disasters can be reduced, by the same action so can vulnerability to the large ones as well. These intermediate events may sometimes be recognized as 'disasters' and sometimes not; sometimes by insiders and sometimes by observing outsiders. In some cases they may be recognized only by insiders who may or may not be able to cope, and not by outsiders who, if they know of the event at all, may regard it as insignificant.

The observation was made more than 20 years ago, that prevailing conditions ('long-term trends') governed the rate and quality of recovery after a natural disaster (Haas et al., 1977). Later, Quarantelli has observed that focusing on the 'causal primary' of disaster impact leads to a misunderstanding of at least some post-disaster problems and leads to a failure to recognize that the most effacious solutions may reside in changes to public policy and in intervention aimed at changing aspects of the social structure (Quarantelli, 1985 quoting Golec, 1980).

More recently, Mitchell et al. (1989) have concluded that hazards come from a variety of sources and that it is therefore desirable to design public policies to take account of that variety and of an inevitable interrelatedness within it. In order to do this, it is necessary to involve not only those whose work focuses on specific hazards, but a broad spectrum of organizations and groups that affect, and are affected by, the many contexts of hazards.

These geographers and sociologists are saying that activities seemingly outside of the academic discipline and managerial sector to do with natural hazards, have probably the greatest significance on their outcome and impact.

A paper of the Natural Hazards Working Group based at the Universities of Colorado and Toronto (Kates, 1970) had earlier observed that 'the real determinants. . . related to natural hazard lie outside the interface' (of the consequences). One example given was the encouragement of cash cropping and prohibitions on migration by colonial administrations in East Africa that had probably intensified the disastrous effects of droughts in the 1930s. In other words, it is most likely to be the policies and activities which seemingly are undertaken without reference to natural hazards, that eventually have the greatest bearing upon them.

The same author also observed that such critical events, seen as important with hindsight, are not easily handled by means of 'adjustments to hazards', regarded as the key to successful ecological interaction between man and nature. Now, with further hindsight and from the viewpoints considered in this book, it seems that 'events such as these' continue to be regarded as critical but that they are still not part of an overall understanding, least of all part of an overall strategy for dealing successfully with natural hazards.

It seems relevant to observe now, that 'adjustments' were a strategy devised in a highly developed and rich country, where there existed options and resources with which to implement adjustments to hazards. What it also seems useful to observe now, is that in poor and less developed countries, whilst on the one hand adjustments were less possible, development as the result of external initiatives and external funding was possible. Here, there may have been the worst of both worlds as far as natural hazards were concerned; development that did not take notice of the potential for natural hazards, and indigenous populations that could not – except by their own cultural and historical skills, which were to become eroded in time by the effects of 'development'. In these contexts and circumstances, changes to prevailing conditions and contexts have been far more significant in the impact of natural hazards than 'adjustments' could ever have been.

It is largely what we understand as development that in its matrix of physical, technical and social change is the incumbent of contexts and prevailing conditions and of the policies and activities ensuing from them. Although it would seem obvious that development practices should have moulded (adjusted) themselves to the eventual manifestation of hazards within them, this has demonstrably not been the case.

Since the early 1970s, the issue of the relationship of disasters and development has been a topic of intermittent writing and discussion, only to fade repeatedly as increased demand for emergency action has focused on necessarily short-term responses.

These conditions may have brought about a disregard of the longer term. The subject of disasters, 'natural' or otherwise, is essentially event driven. No sooner are there indications of consensus, such as it is, than we are pressed to deal with another set of events and circumstances. Nevertheless, natural hazards, as distinct from complex emergencies and conflicts, have not gone away.

This pressure to move on to new circumstances and new issues has exacerbated a failure to retain some basic concepts established in the 1960s and 1970s. This book attempts to recall, reassert and to reapply some of these basic concepts, but with the message now that development is the prime medium of vulnerability and its reduction.

Where development has been considered at all in relation to disaster response, it has been regarded as the long-term objective after the disaster has occurred, relief has arrived, preparedness re-evaluated, rehabilitation achieved and reconstruction commenced. With some modification and long-term thinking, it is conceded, these processes could lead as a sequence into development. Much of the thinking surrounding the relationship between disasters and development has been (and still is) about how post-disaster response can better be made to relate to 'development'. In the meantime, development has run the risk of being interrupted and impeded (or even negated) by disasters and post-disaster responses, but development has been something apart from such inconveniences.

The disaster–development 'continuum' has been the the most recent expression of this concept.

The objectives of development are not an immutable panacea or nirvana, achievable only when all the hurdles have been jumped. Development is (usually) a planned process of change which as the creator of prevailing conditions and contexts within which people live and participate, becomes the framework within which all else happens – planned or unplanned.

As the source of prevailing conditions within which occur environmental hazards from time to time, it follows that one of the roles of development has to be to incorporate adjustments to the likelihood of hazards – both environmental and otherwise. The identification of ways in which development might be modified and 'adjusted' so as best to do this is the purpose of this book.

Much has been written on how disasters impede and negate development, but the key question has to be how development can affect the consequences of disasters. This book examines the requirements for long-term change so that conditions which have become the contexts for catastrophe can be modified. Such an objective is not easily identified, least of all achieved, when popular attention is necessarily focused upon short-term measures for emergency relief and post-disaster assistance, and for the immediate saving of lives which might otherwise be lost.

The argument of this book is that development for vulnerability reduction is not necessarily the last in line of a sequential process that has to wait for a disaster to start it off. Invariably in practice – and contexts of conflict are no exception – necessary immediacy of short-term action takes priority over the longer-term; but if longer term policies were already established, not only would short-term action have a positive context within which to contribute to development, but implementation of policies for vulnerability reduction in development could commence.

The book focuses on those longer-term policies and activities, with respect to those engaged in efforts for the short term and in the hope that recurrence might be reduced by a modified future.

The literature on disaster vulnerability is now extensive; consequently this book is able to take it as read that its theoretical basis is established. Nevertheless, some theorising has been superseded, there are gaps, and there are circumstances for which some of the theorising is not relevant. As always, there are experiences and descriptions which can be added to an increasingly rich spectrum of understanding of relationships with natural hazards. This book may be a small contribution to that process.

It is customary for writers on disasters to commence with catalogues of the enormous scale of disaster impacts and for this to be magnified by aggregation in global terms; e.g. four million deaths in 25 years; global cost of disasters is US$40 billion a year; etc. Such global aggregations have been the basis upon which global organizations and initiatives have been established.

Disaster response has been overly influenced by global comparisons of disaster magnitude, by large organizations to counter them, and with costly technological infrastructure to resist them. Consequently, it has been insufficiently concerned with local experience, context and daily survival.

As extensions of everyday hazardousness, however, small disasters occur for small numbers of people, and news of these never reaches the accountants of the global statistics. They occur in remote places, far from reporting networks, or in small and distant countries, the national populations of which are smaller than the number of casualties in some disasters of international repute. In those places, disasters cannot compete for international headlines.

This book is concerned with those places, where 'disasters' occur repeatedly and are a matter of everyday existence, as they have been since populations began. Its thesis is that indigenous attention to small matters in the face of externally perceived massive catastrophe is sensible and effective; but that popular concern mostly with disasters of globally impressive magnitudes is causing us to forget, to preclude, or to destroy local coping strategies and mechanisms.

Awe of disaster magnitude has brought about a separation of disasters and their management from everyday affairs. 'Disaster management' has been separated from 'development', and vulnerability has been neglected and consequently has been exacerbated. This separation needs to be redressed so that appropriate development can attend to the reduction of vulnerability, a prevailing requirement common to the effects of hazards of all kinds, of natural disasters and of conflicts, both in their duration and in their aftermath.

Conventional development (i.e. economic growth) is not a prerequisite for disaster reduction – which would in itself be development. Appropriate and sustainable development can be tailored to meet basic needs for vulnerability reduction, survival and recovery – and within local environments, capacities and aspirations. Recovery will express sustainability, but measures for the achievement of recovery must themselves be sustainable.

It matters less what kind of disaster is likely, than that socio-economic contexts are able to absorb and to cope with not only their initial impact but their aftermath as well. It is the aftermath which will eventually become the context for the next disaster – of whatever kind.

It is hoped that this book will contribute towards redressing the balance of attention devoted to large-scale disasters of global significance, towards consideration of contexts and prevailing conditions everywhere rather than 'disaster this or disaster that'; and to rural rather than urban locations (because if the rural is put aside it will continue to migrate to the urban), and to a wider understanding of the processes of vulnerability accretion and consequently to its reduction.

When conflict threatens to involve large areas of the African continent, and what was formerly Yugoslavia has been destroyed and divided by ethnic conflict, it might seem academic and self-indulgent to return to the relatively comfortable topic of 'natural' disasters. Before the International Decade for Natural Disaster Reduction was half spent, the 'disasters world' had changed immeasurably – yet natural disasters have not gone away.

As the book attempts to show, however briefly, there is commonality between a kind of development that would assist a prevention of conflict and one that could assist a reduction of natural disasters. Were a socio-economic development programme devised to serve equitably the reduction of natural disasters, it could be made to serve the objective of conflict reduction as well.

Humanitarian assistance has now grown at a rate out of all proportion to development assistance; indeed, development budgets have declined as budgets for short-term humanitarian assistance have increased. What was once called 'disaster relief' has grown into a high profile mega-industry with the popular appeal that many mega-industries might envy – or abhor. Development aid continues, but almost as a poor cousin – just as 20 years ago or less, disaster relief was the poor cousin to development.

It is the purpose of this book to try to show that, in spite of the very real need that does exist for emergency assistance, the reality is that it is serving the symptoms and not attending to the causes. In spite of the mega-industry of humanitarian assistance, the causes of vulnerability to hazards of all kinds are being left to ferment unattended.

Most mega-industries would be gratified by such a growth in their 'market-place', but it is to be hoped that humanitarianism will prevail in recognizing that such growth would be at enormous long-term human cost – and that something must be done to reduce the need, not only to respond to it repeatedly. This book is addressed towards that purpose.

Even in a little thing
(A leaf, a child's hand, a star's flicker)
I shall find a song worth singing
If my eyes are wide, and sleep not.

From a *Gilbertese Song* (now Kiribati)

Translated by Arthur Grimble
in
Return to the Islands
John Murray 1957.

PART ONE
VULNERABILITY

The same geological processes that can create the beauty and fertility of an island can lay waste a third of that island. . . These forces are original and permanent. They form a Natural Presence, which is co-ordinated with our own presence. . . And nothing is more basic to our existence as humans than the undertaking of our relationship with this Presence. It is a way of defining ourselves.

The Volcano Suite: A Souvenir Series of Five Poems by Shake Keane, Kingstown, St Vincent. 1979.

1

The meaning of vulnerability

National and local vulnerability

NATURAL DISASTERS HAVE rightly been adopted as a world problem, but the response to a problem perceived to be global has in many cases been the application of globally assumed and generalized 'solutions'.

It is an indisputable fact that some natural disasters do inflict colossal losses from time to time. The volcanic explosion and eruption of Mt Krakatoa in 1883, in what is now Indonesia, killed 36 000 people as a result of the consequent *tsunamis*. The earthquake of 1923 in Japan killed 156 000 people, many in fires that ensued. The earthquake of 1970 in Peru destroyed the city of Hauraz and caused the side of Mt Huascaran to descend upon Hauraz and to wipe out the entire town of Yungay; 22 000 people died. In the same year, 250 000 people died in Bangladesh as the result of one tropical cyclone; in another in 1991, 130 000 people died.

Preoccupation with individual disasters of largest magnitudes of death and destruction is, however, misleading; strategies are required in all places and for hazards of every type and every magnitude, and in a wide variety of conditions. If the impact of lesser and often unreported disasters were to be aggregated, the sum would probably be as great as any single large disaster in terms of people affected, homes destroyed, or economic loss. This viewpoint can alternatively be expressed by reference to disasters in small countries, the total national populations of which sometimes do not amount to the numbers of victims of some large disasters in sub-continental countries. Many of these disasters are therefore seemingly insignificant by any global comparison of magnitude, the number of people affected each time being small by comparison.

The *proportional* impacts, however, are far greater upon small countries and small places. The tropical cyclone which swept into the Indian state of Andhra Pradesh in 1974 rendered two million people homeless, 4 per cent of the state population and less than 1 per cent of India's national population. When, in 1972, 120 000 people were made homeless by hurricane Bebe in Fiji, they represented more than one-fifth of the national population. Eighty per cent of Dominica's housing stock was destroyed by hurricane Allen in 1980. Twenty-two per cent of Tonga's housing stock was destroyed and 50 per cent of the national population were made homeless by hurricane Isaac in 1982 (Lewis, 1989b). Thirty-four per cent of Jamaica's population were made homeless by hurricane Gilbert in 1988 (Lewis, 1991b). Similar proportional assessments can be made for hurricane Hugo in 1989, hurricane Andrew in 1992 and for many more hurricanes both before and since.

Similarly significant proportions of island populations, whether in the Caribbean, South Pacific or anywhere else, become victims – especially of tropical cyclones. Hurricane Georges in 1998, subsumed Antigua, Nevis, St Kitts and

St Croix, before traversing the lengths of Puerto Rico, the Dominican Republic and most of Cuba and entering the southern states of the USA.

It is always *proportional* impacts which count for more than expressions of disaster magnitude. Comparison of proportional impacts at national and state levels can usefully be applied intranationally between administrative units and communities (see also Case-study IV). Analyses of casualties in comparably affected townships and districts of the Guatemala earthquake of 1976, hurricane David of 1979 in the Dominican Republic, and the Algerian El Asnam earthquake of 1980, all lead to similar conclusions. The smallest administrative units suffer highest proportional impacts, which are invariably rural rather than urban (Lewis, 1980).

In comparisons of domestic impact, it is the poorest who may sustain absolute impact and lose all they have; the proportional impact upon the comparatively well off is often less; and though the rich may lose quantitatively more, they have more in reserve upon which to survive.

Globalized comparisons of aggregated disaster magnitudes and assumptions of generalized vulnerability, inadvertently obscure some important considerations:

• Categorization of disasters by magnitude reflects a remote and privileged comparative view which tends to exclude disasters of a lesser degree, even though locally these may represent catastrophic national and local impacts. Magnitude thus tends to obscure the concepts of local experience, understanding, context and capacity.
• Vulnerability is similarly and essentially a local condition, and its understanding invariably ensues from local context, experience and analysis. Though vulnerability pervades in a global sense, understanding of its causes and characteristics is not assisted by generalized 'globalization' (Lewis, 1988a).

Global policies for disaster reduction, such as they are, have to be augmented by local programmes. These programmes will be far less dramatic than the disasters they seek to reduce. They are unlikely in themselves to create news headlines, but in their multiplicity and integration they will do more to reduce those disasters of greatest local significance and impact.

The meaning of vulnerability

Vulnerability is the degree of susceptibility to a natural hazard. The concept of vulnerability is a significant contribution to our understanding of natural disasters (Baird *et al.*, 1975; Maskrey, 1989; Hossain *et al.*, 1992; Winchester, 1992; Blaikie *et al.*, 1994; Hewitt, 1997). The vulnerable state of populations and settlements is as much a contributor to the cause of 'natural' disasters as are the physical phenomena with which they are associated. What are called 'earthquakes' and 'hurricanes' are the natural forces; what are seen afterwards are the results of the impact of those forces on human settlements. The degree of susceptibility to damage, destruction and death in those settlements is conditioned by the decisions and actions of society over time. This means that there exists a social, institutional and political responsibility for a major proportion of those causes, and for making efforts to remove, alleviate or to defend against them.

Vulnerability is the product of sets of prevailing conditions within which disasters may occur. Vulnerability has to be addressed therefore, not only by post-disaster concern and response, but as a part of the day-to-day management of change – whether or not that change is called development. The pervasive conditions of vulnerability cannot be allocated as the responsibility of one desk or department; they are the prerogative of all desks and departments for all kinds of business and activity, both policy and practice. In these policies and activities may otherwise ferment the causative conditions for disaster, which so often go unrecognized and unattended until disaster has happened.

Separation of post-disaster responses from pre-disaster contexts inhibits the creation of necessarily wider strategies. The policies and activities of some sectors may even inadvertently contribute to the causes of 'natural' disasters which the relief sector of the same government is then called upon to attend to and to pay for (Lewis, 1984c).

It is, however, simplistic to assume that people are the perpetrators of the disasters of which they themselves come to be the victims. More often than not, the disasters affecting one group of people are caused by another group in the same or in a different place, and at the same or at a different time.

Victims are often said to have been 'in the wrong place at the wrong time'. Vulnerability, as the degree of susceptibility, may be interpreted as simply a matter of location or place – and some places are more vulnerable than others. There is much more however, to the understanding of vulnerable conditions than their physical recognition and identification or, for that matter, than in physical resistance to natural forces in constructional and infrastructural technology. Social and political issues may have had a greater part to play.

Vulnerability, similarly, is often interpreted as a physical state of exposure related to location and quality of construction. Vulnerability becomes manifest when locations and constructions are seriously affected by storm, flooding or landslide for example, and when some poorer areas of construction sustain more damage than others. The external pressures over which victims may have no control, which prescribe such locations, and the lack of options as to where they live or their quality of life, as well as construction, are examples of how vulnerability is related to overall policies and activities in the control of others. To seek modifications through development programmes and projects to avoid or reduce such pressures, is one effective objective of vulnerability reduction in development.

The recognition and identification of locationally or socially vulnerable sectors of populations is itself only an indicator of the *processes* that have brought about those conditions. They are the visible and tangible manifestation on the surface, so to speak, of invisible and intangible social, economic and political undercurrents; and they will have been active remotely and indigenously, contemporarily and historically.

To the geographer Kenneth Hewitt (1983), vulnerability is very much the product of an all-too-technocratic posture by man in regard to his environment. 'Disaster prevention' is an overt expression of the technocratic framework and as such cannot hope to achieve change in those processes that are the causes of vulnerability.

It may be accepted that 'hazard' strictly speaking, refers to the *potential* for damage that exists only in the presence of a *vulnerable* human community. . . There are natural forces and some damages in most disasters that lie beyond all reasonable measures any society could make to avoid them. . . most of them would not be disasters, and many of the damages would not occur except as a direct result of characteristic and vulnerable human settlements. Those developments record mainly the mismatch between the requirements of sensitive, secure environmental relations at the local and regional levels . . . and the demands of those extensive geographies of power and economy with which technocratic strategies have grown up, and mainly serve. . .

. . .hazards research has invented its problem field to suit its convenience. It does not reflect upon the extent to which the institutions it serves – the societies that have made such technocratic authority possible – could be part of the problem. It does not reflect upon the flaws in itself, except in relation to what is deemed sophisticated in the current fashion of the scientific community. It gathers data about people at risk, but may not engage in dialogue with them. . .

. . .the restorations of productivity and reimposing of 'normal' relations become the main prescriptions of crisis management, relief and reconstruction. The ability to predict or contain natural processes in a technocratic framework becomes the main goal for disaster prevention. Now, I question whether this recognises some major, indeed *the* major, ingredients of disaster.

Winchester (1986) recognized with scepticism an over-emphasis on the technical aspects of natural hazards that served to exclude a wider range of solutions to vulnerability. There was a need to prevent people from becoming vulnerable, rather than to aim to mitigate the effects upon them of the natural hazards to which they were susceptible. This he saw as an 'alternative view' to the more popular, more convenient and more powerful 'technocratic view'. These views are more deeply examined within rural communities as a detailed case-study in a part of Andhra Pradesh, India (Winchester, 1992).

The sociologist Enrico Quarantelli (1986) suggests certain advantages that accrue from seeing disasters as social rather than physical happenings:

All disasters are always primarily the result of human actions. A disaster is not a physical happening, it is a social event. . . disasters are in one sense the manifestation of the vulnerabilities of a social system (and) prime attention should be given to doing something about such vulnerabilities . . .

. . .thinking of disasters as social rather than physical happenings [means first] that emphasis comes to be on internal rather than external factors. . .

6

[and second] that disasters as social phenomena [allow] them to be seen as something which can be reacted to as part of ongoing policies and programmes of national and social developments – which could reduce societal vulnerabilities in the first place. Activities of a developmental nature then can be seen as an integral part of disaster prevention and mitigation. There is a tendency for the latter to be treated as a separate sphere of action and responsibility.

Vulnerability as the crucial issue was earlier identified by the Disaster Research Unit at the University of Bradford (e.g. Westgate and O'Keefe, 1976) and was continued in work at the University of Bath in Caribbean and Indonesian field research which geographically exposed social vulnerability as the creation of commercial decision makers (Jeffery, 1982). This work pointed to the need for a broader framework for disaster analysis and for strategies to reduce vulnerability to be an integral part of long-term development. Field study in the South Pacific placed problems caused by hurricanes firmly into the context of everyday shortcomings, which hurricanes exposed and exacerbated. Issues relating to local water and food supply, building construction, development and disaster relief, illustrated how socio-economic, cultural and political contexts could be moulded positively to reduce, or negatively to increase, the impact of frequent hurricanes (Lewis: 1981b; and see Case-study II).

Carlo Pelanda (1981) has concluded that 'the type of organizational state of a (sub)system generates the preconditions for any sort of destruction, natural or man-made', that 'sudden, rare, random, unexpected, destructive events are only symptoms of what we do not know, or of what we are not able, or we do not want to organize' and that 'disaster is the actualization of social vulnerability'.

Vulnerability is often most obviously understood as being related to built structures and the damage that they sustain (Snarr and Brown, 1979). Damage to buildings, as domestic, social and commercial infrastructure, is a crucial and specialist aspect of vulnerability and one that may usefully be used as an indicator of vulnerability in other sectors (see Part 3).

It is easier for most people, including some policy-makers, to envisage a state of affairs in physical terms, especially as we see the effects of natural disasters, or have them presented to us, in such strong images of physical damage and destruction. Entire programmes and budgets have been expended on resistance to the physical impact of natural hazards. To do so without an understanding of the social, economic and political contexts as potential contributors to the causes of vulnerability in human settlements, is similar to painting over the place where water comes in without attending to the reasons for its ingress.

In buildings, the physical aspects of quality of construction, appropriateness of form and material, age and maintenance are easily identified, but the greater value of these and other characteristics is as metaphors for other conditions of a non-material kind. These metaphors are similarly applicable to the social services of health and education, for example, to agriculture and marketing, and to institutional organization and management in all sectors.

Damage to buildings has been used in Peru as the indicator of ground conditions for seismic microzoning (Kuroiwa, 1982) and has been the key indicator of socio-economic level in urban/rural comparative assessments of vulnerability to cyclones in Sri Lanka (Lewis, 1984b; and see Case-study IV). That vulnerability is visibly and more obviously related to built structures and the damage that they sustain, is one reason why the less obvious aspects of vulnerability may go unattended.

Vulnerability is compounded by the degree to which a community is at risk and to which socio-economic and socio-political factors affect the community's capacity to absorb and to recover; by availability of and accessibility to resources, and on the personal and domestic level, by 'defencelessness and the inability to cope with risk, shocks and stress' (Winchester, 1992). Events which may contribute to that inability may accrue over time as the result of a variety of external factors such as economy, loss of land or crops, ill-health, or the death of family members or working partners, all of which could be caused by disasters as well as contribute to the overall effect of them. Thus, vulnerability as the condition of exposure to the initial impact and its immediate effects is only a part of the overall, pervasive and negative condition of vulnerability.

Vulnerability and risk

The assessment of risk tends to focus upon 'elements at risk' in aiming at a quantifiable product of that risk; for example, the numbers and kinds of buildings destroyed, the amounts of agricultural crops lost, etc. Risk assessment also focuses upon the source or origin of the hazard, so as to determine elements at risk to each hazard. Risk assessment looks at the origin of hazard and the effects of that hazard upon elements at risk. Above all, risk is concerned with the *product* of hazard.

Consideration of vulnerability, on the other hand, looks at the *processes* at work between the two factors of hazard and risk. It reverses the conventional approach, and focuses upon the location and condition of the element at risk and reasons for that location and condition of infrastructure, construction, community, dwelling, population or person. Focus upon source or origin of hazards maintains a difference between them; whereas focus upon vulnerability maintains a similarity of sustained effects as a result of hazards experience. By attending to vulnerability, the effects of all potential hazards can be accommodated to some degree – from the point of view of the victim's potential to survive and to recover. Measures to decrease vulnerability are partially integrative with normal collective conditions, small-scale, and individually are less costly and more achievable.

Risk is the product of hazard and vulnerability. Hazard is the potentially damaging (natural or man-made) phenomenon; vulnerability is the degree of susceptibility to a hazard (van Essche, 1986). Risk is thus a statistical probability of damage to a particular element which is said to be 'at risk' from a particular source or origin of hazard. Damage or loss is expressed as of a certain magnitude and as occurring in a particular location or area and during a particular period of time (Lewis, 1987c). The longer the time period, the greater is the possible magnitude.

Focus on fearful risks of large magnitude clouds our perception of a reality which in the event may be lesser or partial. In other words, being faced with the concept of a large risk makes it more difficult to conceive realistically of risk probabilities which may be smaller, more normal, and more frequent.

Risk, as the probable manifestation of a hazard, focuses on the potential of occurrence and not on the period of time for which that occurrence is assessed; but risk is a prevailing condition or a state of affairs, not the event itself which causes the damage. Changes will inevitably occur during the time period, which will affect the degree of susceptibility to hazard and the amount of potential damage sustained. Thus, where an event occurs as the manifestation of a risk (in accordance or not with its actuarial probability), the effects of that occurrence may be different at the end of the period than they would be for a similar event at the beginning of the period.

There may be more than one source of risk, numerous elements at risk, numerous changes which take place within them, different rates of change and thus variable conditions of susceptibility. The overall condition that results is vulnerability. Risk is static and hypothetical, (though reassessable from period to period of time) but vulnerability is accretive, morphological and has a reality applicable to any hazard. It is not dependent upon, or applicable to, only specified sources of hazard.

Therefore, it is not only risk that needs to be considered; it is also vulnerability. Vulnerability is actual; risk is actuarial. Whether or not risk is appropriately assessed, vulnerability will accrue or change. Conversely, changes to susceptibility, either planned or fortuitous, may cause vulnerability to increase, diminish or to be held in check – independently of assessments of risk.

Risk is an assumed totality of an event brought about by forces over which we have no control (or over which we have lost control). Counter-measures against risk are therefore conceived as separate from normal activities, powerful and massive, and therefore usually of high cost. Vulnerability, on the other hand, is accretive and aggregative, and has to do with susceptibilities resulting from ageing, weakening, limits to options, economic and social level, degree of integration and access to resources and services, and with the reasons why these either are or are not positive (Pelanda, 1981).

Perception of total risk may be so intensive, painful and fearful, as to inhibit the very measures that might otherwise bring about its mitigation. Analyses and assessments of vulnerability, on the other hand, relate to that variety of susceptible elements which contribute to the total vulnerable condition; as such, they are more comfortably, more realistically and more constructively considered and implemented.

A number of issues emerge from these concepts. First, the condition of risk is a construct and calculation for which certain static values are necessary. Fluid or changing situations do not lend themselves to simple mathematical 'certainties'.

Second, risk has been conceived as the product of hazard and vulnerability, in the convenient formula:

$$Hazard \times Vulnerability = Risk$$

For these purposes, the hazard has to be identified and specific, and be of a given magnitude; and vulnerability has to be a static and given degree of susceptibility. As a component in the equation, vulnerability is made to be secondary to the consequence of the equation, which is risk.

Third, risk is a prevailing state of affairs; it is not the real source of damage or loss. Focus on risk of a given magnitude may cloud our perception of a reality which might in fact be lesser or partial. In other words, being faced with the concept of a large risk may make it more difficult to conceive of probabilities that may be smaller, more normal and more frequent – and of what to do about them.

The consequences of this interpretation of magnitude are far reaching. Where resources are scarce, it is the smaller, normal, and more frequent risks that we are able to take measures against but, clouded by our focus on the risks of greater and more fearful magnitude, we might give up and in the face of these – and do nothing. We shall thus sustain loss from more frequent risks when we could have protected ourselves, and subsequent vulnerability to the next disaster – that would otherwise have been of limited extent – will be increased and be more widespread.

We can rarely forecast our position relative to the centre of the source of risk. Shall we be annihilated outright or shall we be on the periphery and experience that risk as if it were a minor event? The taking of seemingly minor measures in the face of apparently colossal risk is therefore not fallacious (Lewis, 1987a).

Risk is a prevailing state of affairs conceived largely for policy formulation and managerial decision-making. As such, it is a construct by 'outsiders' of potential events in other places. Though reassessable from time to time, it is essentially and necessarily a static concept.

Risk is an aggregated state of affairs, combining and amalgamating conditions in a number and a variety of places or separating 'elements at risk' from everything else. Governments and institutions, as outsiders to the locality, may be interested in risk as a value from other places. Vulnerability, on the other hand, is a prevailing condition or interrelated set of conditions. Though assessable by 'outsiders', it is as a prevailing condition inevitably an 'insider' experience. To communities and others undertaking localized activities, vulnerability is the more relevant condition, measures of which they may perceive for themselves (Rahman, 1991).

Risks of varying degree can be mapped conveniently in contour and colour, superimposed upon habitation and infrastructure. The origins of risks, or causes of exposure to them in terms of where and upon whom they might impinge and what effects they might have, make mapping a much more complex undertaking, though – where practicable – a more realistic one is the commencement of an expression of vulnerability (see also Chapter 2).

The experience of hazards can be considered in two stages. The first stage relates to initial impact, when vulnerability may be conditioned by location, density and distribution of population, age of people and of structures, and technical and social capacity for resistance or protection. The second stage relates to the aftermath and to the capacity of survivors to continue to survive in the longer term. Sustained survival requires an effective culture or infrastructure of indigenous assistance, resources and social services. The greater the area and magnitude of hazard impact, the less available, effective or accessible will be external assistance – and therefore the more essential are indigenous resources and facilities.

2

The observation, perception and identification of vulnerability

UNDERSTANDING ABOUT VULNERABILITY has had the effect of taking the study of natural disasters by those who observed, recorded and compared them from afar as 'outsiders' (Lewis, 1977) to a viewpoint more closely related to the experience of natural hazards within their social and economic contexts.

Observation of vulnerability

Not only were other people's natural disasters observed and described by outsiders – each outsider was of a different speciality or discipline. Seismology and meteorology for example, divided the study of earthquakes and tropical cyclones between them, and they remained disciplinarily divided until study of the experience of both in the same place conveyed a semblance of the reality of people's vulnerability. It was not until *The Hazardousness of a Place* (Hewitt and Burton, 1971) examined the experience in London, Ontario, of a cumulative range of hazards of different kinds over time, that this significant shift in disaster studies commenced.

The shift was into the study of hazards in a context of human settlement, and towards the disasters that ensued, as distinct from the study of the phenomena that were assumed from afar to have caused them. The difference was the incorporation of people, their culture, communities and society, and their physical and social infrastructures, into the study and analysis of disaster impact. Simultaneously, development of the concept of vulnerability recognized the participation of people and communities in its ecological process as well as in its condition (Kates, 1970).

Perception of vulnerability

The changes that eventually came about brought us nearer to the reality of natural disasters as they are experienced, as distinct from as they are studied. Instead of discrete events for monodisciplinary study, we are now concerned with the manifestation and exposure of hazardousness in contexts of vulnerability. Extremes of these conditions occur as an irregular series over time, each of differing origins (agents), magnitudes and epicentres, and each having a bearing on vulnerability to the next – of the same or different kind.

It was possible at the end of the nineteenth century to know of earthquakes in any part of the world by the process of remote sensing established in 1889 by Gutenberg and Richter (Lewis, 1977). By the same token, it was rarely possible to know their effects. Earthquakes could be recorded and studied without their impacts on people and their social and economic contexts being known. The rare exceptions demonstrated the rule; Angenheister's description (1921) of the effects of an earthquake on an island in Tonga are graphic (see Case-study II).

By contrast, information regarding the track of tropical cyclones relied upon separate reporting of experience of them, usually from ships at sea. It was often only after a considerable time that the sequence of experiences could be established, from which the track of cyclone behaviour became known. The 'calamitous visitation which swept the greater part of the Bahama Islands' in October 1866 was not finally fully reported until March 1868 after 34 separate reportings of it from on land and at sea (Rawson, 1868; in Lewis, 1977). 'Remote sensing' of tropical cyclones came much later, with its retrogressive contribution to knowledge of the experience of them on the ground.

The origins of the remote recording of what inevitably were 'disasters' for the people who experienced them are not simply matters of history. Detachment from the impact of natural disasters has been maintained in many research undertakings concerning their incidence. The magnitudes of distant events have been recorded and compared; not the local impact in proportion to geographical contexts of land or population size or of economic impact in comparison to production data. This detachment persists, even though it was first commented on 20 years ago (Lewis, 1977).

The result of this is that the 'large' disasters attract world attention, but 'small' disasters rarely do; even though 'small' ones in small places may be of catastrophic proportional impact. Similarly, local disasters anywhere are discounted in these comparisons, though small local disasters are more frequent and more indicative of a normal hazardousness.

Bangladesh is infamous for having experienced some of the most cataclysmic natural disasters (since one of the bloodiest wars of independence) in the form of frequent floods, and in tropical cyclones in which thousands of people have lost their lives. These natural disasters rightly also hit the world's media headlines. What do not hit the international media headlines are, for example, the constant cases of river bank erosion which are regularly the cause of homelessness, landlessness and consequent migration for many thousands of people every year (Rahman, 1991).

These comparisons have a crucial bearing, not only upon understanding of the relationship between context and vulnerability, but on what to do so that disasters can be reduced. The identification of what kind of development is required for disaster-prone places is the objective of Part 3.

On field missions, invariably in the wake of reported nationally and internationally significant natural disasters, there is revealed evidence of other recent disasters of local comparable impact, or there are local conditions and issues of equal or even greater importance that the disaster has exacerbated or compounded. It has often been difficult for an assessor from outside the affected area to distinguish in field analysis between damage caused in the disaster of recent report and damage caused by earlier disasters of the same or different kinds left unrepaired from years before. At once one natural disaster is in its local context, an environmental and ecological context formed partially out of the effects of intermittent and recurrent disasters of all kinds, themselves inter-related and inextricably further interrelated in their manifestation with the totality of socio-economic, cultural, political and natural environments.

Vulnerability as a condition accrues as a variety of processes over periods of time. For analysis of vulnerability not to take account of its causative processes, or for it to assume that vulnerability now is a static product of expired processes, is a seriously incomplete approach.

Identification of vulnerability

Natural disasters occur only where natural extremes of the environment meet a human community that is vulnerable. The condition of vulnerability comprises numerous interrelated components and factors; for example, location, age, eco-nomic levels, social divisions, administrative and political integration, accessi-bility of resources and services. These factors have a bearing upon which activities are undertaken within the community and by whom, to what degree they impinge upon socio-economic vulnerability and for whom, what options there are for change and modification, and ultimately whether these options allow environmental conservation or whether they force or perpetrate environ-mental degradation and the removal of natural environmentally protective conditions.

Some communities and some sectors of communities therefore become more vulnerable than others; vulnerability being an expression of changing social and economic conditions in their relationship to the nature of the hazards to which they are exposed. It is within these complex and interrelated processes that vulnerability is created. Vulnerability is therefore not static; vulnerability is dynamic, evolutionary and accretive.

Although vulnerable conditions are plain to observe in the aftermath of natural disasters, and are assessable and measurable in post-disaster investiga-tions, how can vulnerability be identified and assessed *before* disasters take place? This requires a return to consideration of socio-economic vulnerability as it affects levels and sectors of society. Vulnerability can be considered relative to each level and sector, and the relationships within and between different levels and sectors can be examined.

In the field, it is possible to apply a vulnerability matrix or model at any level or in any sector. In reality, however, indicators of vulnerability per community or region are appropriate and accept some homogenization of vulnerability values. In other words, for working purposes it may not be necessary to know the vulnerability value per family, when families share a common vulnerability reflected in indicators of economic or social values. These indicators may be to do with location, house type or construction and/or family size, for example – indicators of a general vulnerability for an area as well as vulnerable elements themselves. There are other, more remote and seemingly less relevant, socio-economic indicators that can be applied in socio-economic assessments at local levels. Examples of these are described and utilized in the method used for field assessments in Sri Lanka (see Case-study IV). Significantly, housing was an indicator of socio-economic level as well as itself being subject to damage and destruction.

For most working purposes it has to be assumed that physical vulnerability is common throughout a community, area or region. In these cases, simple distribution of population becomes the indicator of vulnerability. Detailed vulnerability assessment in the field is a time-consuming task for which basic data may not be accurate or even available. It is dangerously inappropriate however, to use population as an indicator of vulnerability for the purposes of establishing priorities of need; small populations would indicate low vulnerability, leaving those small and vulnerable populations in a position of low priority for assistance as a result. It is very often small populations that have chosen, or have been obliged to choose, the most vulnerable locations for their settlements. Should these small communities be denied support simply because they are small or non-conforming? Furthermore, the use of population densities as indicators of vulnerability has to take account of population movement and migration, both permanent and seasonal (see below).

Preferably, of course, what are required are overlay maps or geographic information systems that will facilitate the plotting of contours of hazard and risk together with mapped indicators of socio-economic vulnerability; but unless risk contours have already been prepared (e.g. Fournier d'Albe, 1976) the time and cost incurred may make this option impracticable.

Local vulnerability is being assessed in part with the participation of local communities. An example of what might be called micro-vulnerability assessment, has been a result of rural village participation in the Philippines (Hall, 1996, 1997). With the help of a large three-dimensional village map, made in local materials for the purpose, villagers were asked to indicate the houses they considered to be vulnerable, to which hazards, and for what reasons. Flooding, landslide and typhoon (tropical cyclone) potential were taken into account by villagers, as were building maintenance, ownership and tenancy, recent settlement and migration, and livestock security.

However, both exogenous observation and indigenous local perception, on their own, may be insufficient. Local perception of hazard may not be total; activities outside of the locality may have created or exacerbated hazards, the effects of which are as yet outside local experience. Local activities may therefore be undertaken in ignorance of their effects upon vulnerability to these as yet unknown hazards. More simply, there may not have been a serious earthquake within living memory and beyond, but that may not mean that the locality is not earthquake prone. Both vulnerability and perception are dynamic and shifting; assessment of either at a particular point in time may not conclusively convey the total picture at that time – nor for any other time in the future.

Real and comprehensive vulnerability assessment has surely to be a fusion of local and regional/national, or of micro- and macro-vulnerability assessments – even though the latter on its own is a participatorily poor and culturally impoverished substitute too easily adopted for expediency.

In Algeria, data were not sufficient to permit localized comparison between population distribution and seismicity. A homogeneous earthquake hazard had to be assumed until data became available (Lewis, 1981a). In the interim and in a temporary assumption of an evenly distributed earthquake hazardousness, the distribution of population, housing and infrastructure was used as the primary indicator for an analysis of vulnerability. A similar working method was adopted in Tonga (Lewis, 1978).

Working with data in Bangladesh, of the smallest administrative areas (*mauzas*) and in a temporarily assumed equal cyclone hazardousness for the coastal area, population density has inevitably been used as the indicator of vulnerability to cyclones, thus homogenizing the variations of location and socio-economic level within each *mauza*.

Risk could be assessed simply by assessing the hazard as water depth (homogenized) and multiplied by population density. This however, would be an 'outsider' assessment. Further screening would introduce local characteristics, such as location (island/isolated mainland/mainland); level of existing protection (none/partial/inside low polders/inside deteriorated polders/straddling embankments/below embankments); existing cyclone shelters available (adequate/not adequate) (taken from Pitman, 1997).

Without such further screening, low population numbers would be assessable as low risk because of low numbers – even in a six-metre storm-surge! There are numerous 'small populations' at such exposure. 'Locational vulnerability' overwhelms and would wipe out minor variations of 'vocational' vulnerability – such as economic level, if such data were available or obtainable at *mauza* level. On the other hand, situations could occur where socio-economic vulnerability might be highest where locational vulnerability was low. Each situation requires field assessment, if generalized 'outsider' vulnerability assessments, based only upon population density, are not to prevail.

Movement of populations

Movement of populations over time, therefore, has a significant bearing upon vulnerability assessed in this way, both because movement shifts degrees of indication from place to place and because in the shifting, actual locational vulnerability of populations changes with change of location, as does the vulnerability of groups and families which constitute larger populations.

Induced, coerced, forced, or spontaneous population movements have frequently been a direct cause of vulnerability, removing people from their accustomed resource base and creating conditions of dependency, which exacerbates disaster – and which disaster exacerbates. Changes in the location of population may, over time, bring about more direct consequences of vulnerability; deaths, injury, housing-destruction and homelessness may be increased by movement of populations into more hazardous environments and/or more fragile socio-economic contexts. Population movement therefore may be the cause of two kinds of vulnerability; the first, due to communities having been separated from their resource base, with reduced capacity for self-reliance and more dependent on external assistance for support – in 'normal' times and therefore especially after natural disasters; and the second, populations caused, knowingly or unknowingly, to inhabit a more hazardous location, thus placing more people at risk. These two sets of vulnerability exacerbation could become combined – to create a third compounded case of vulnerability.

These processes may be further compounded. Deprivation may be the cause of migration; migration may cause overpopulation of a resource base and create deprivation. There are likely to be socio-economic causes of the migration which exacerbates disaster, which may itself, in turn, be the cause or motive for further population movement. Analysis which takes simple population increase, or changes in population distribution, as the only basic factor of vulnerability to disaster will be incomplete and an inadequate expression of vulnerability as a complex and dynamic process.

A comprehensive strategy (see Part 3) is required in which all components of a programme for disaster reduction are simultaneously set in motion. A policy of population reduction may be one multi-sectoral component for disaster reduction, but whilst little comprehensive strategy is being effected, disasters themselves will remain for many people a significant factor of the hazardous uncertainty against which large families are the only accessible forms of insurance.

Urban centres are often assumed to be subject to the greatest disaster impact; but these are largely the result of mono-disciplinary judgements, encouraged by information sources unable or unwilling to reach rural hinterlands.

The normally fragile relationship of a population with its environment and climate has frequently been upset by the superimposition of disruptive policies. A community removed from its traditional resource base is made more vulnerable to repeated or alternative hazards and, moreover, may be unable to exercise effectively its traditional responses. Similar conditions of high rural vulnerability and exposure have been identified by field research, for example in Sri Lanka, the Dominican Republic, Martinique, and Indonesia (Lewis, 1981c; Jeffery, 1981a and b).

The tropical cyclone which destroyed the town of Batticaloa on Sri Lanka's east coast in 1978 traversed the country and caused further damage to inland areas, where rural populations typically predominate. Field analysis (Lewis, 1984b; and Case-study IV) showed much larger numbers and proportions of destruction and damage in rural areas than in urban areas*; but it was the spectacular, concentrated and easily accessible damage in Batticaloa which became the focus of national and international assistance.

In many cities and urban areas, however, evident vulnerability is largely a *fait accompli* in history. It is building damage which makes manifest the earthquake and it is the collapse of buildings which in turn causes deaths and injuries. The age, condition and strength of buildings is therefore an indicator of vulnerability within cities – the design and construction of recent buildings being as significant as the age and condition of old ones. Where older housing stock is occupied, and over-occupied, by poor populations, buildings and social vulnerability are closely

---

* Percentages of housing destruction in the Sri Lanka cyclone of 1978 were highest in low-density rural areas and lowest in high-density urban areas. (Lewis, 1981c; see also Case-study IV).

linked (Lewis, 1987b). When buildings are old to the point of dilapidation, it is the poor who are either obliged, or who choose (without accessible options), to occupy them; usually because they are cheap to rent (or rent free). Because rents are low or non-existent, the buildings are not maintained. This kind of situation was exposed by the 1985 Mexico City earthquake (ECLAC, 1985). Earlier tremors and small earthquakes may also have contributed to a weakening deterioration of buildings, which are thus made more vulnerable to destruction in the future.

Social vulnerability within urban areas may more obviously be expressed by the location, as well as the quality, of buildings. Guatemala's capital was moved to its present site after the former capital was destroyed in 1773, after which it was virtually destroyed again in 1917. The 1976 earthquake killed 22 000 people and destroyed much squatter housing which had been built on the sides of the ravines that surround the city. Buildings in the city centre were relatively undamaged. The majority of casualties and homeless were, however, in rural areas (Olsen and Olsen, 1977).

Historical studies

Catastrophic unpredictability of earthquakes and tropical cyclones often obscures perception of their recurring frequency. The consequently pervasive hazardous context of human vulnerability can thus come to be disregarded. Historical studies which can expose the crucial characteristic of recurrence often do so only for one disaster type, rather than including all kinds of disasters that have occurred in one particular location. This has often led to an unrealistic and academic separation of events, divorced from their contexts and unable to portray the interactions of environmental reality.

The aftermath of one disaster becomes the vulnerable context for another of the same or of a different kind. Institutional, physical and social structures can be strained or weakened by one event and be unable to resist or counter the next. Alternatively, over-response to one thing may inadvertently create a vulnerability to another. Institutionally there are rarely such linkages, either within the vulnerable area or within assistance organizations. Decisions for development are often taken with disregard for hazardous environments, by a department separated from the management of disaster relief and rehabilitation (see Part 3; and for recurrence of a variety of disasters, see Case-studies II and III).

Small countries and island states

Island countries and countries of islands have, in their relative smallness, an extraordinary vulnerability. Tropical cyclones in their destructive power can engulf entire island groups and cause devastation on a proportional scale unknown in larger and sub-continental countries. At the same time, their diverse and scattered smallness, in archipelagos for example, has special implications for administration and management which provide both constraints and opportunities for development strategy.

In addition to the island states (and territories, protectorates and dependencies), are those islands and island groups which are parts of continental states, such as the Lacadive, Nicobar and Andaman Islands of India, and islands off Bangladesh, Honduras, Kenya, Malaysia, Sierra Leone and Tanzania, for example (and see Case-study V).

The characteristics of place have a significant bearing on the identification of development strategies anywhere – economically, socially and culturally. The place will have had its influence upon local culture and vice versa. Local and regional, popular and official perceptions and analysis of them, are thus necessary for understanding local and regional influences upon vulnerability and strategies to counter them – whether in the aftermath of one disaster, or contexts of vulnerability to the next.

In the shifting interplay of contributing factors of vulnerability, what happens in islands will be much the same as what could happen at local level anywhere. The difference is that islands are immediately 'local level' and appropriately identified strategies are implementable there, being manageable and small scale. Islands offshore of a mainland or continental country especially require processes of development designed on their behalf, and with their participation, rather than to share only in prevailing national programmes.

Political marginalization is the overriding risk shared by all islands and island states. Island states have been able to counter this internationally to some extent by the formation of the Association of Small Island States (AOSIS) and subsequent groupings. Islands offshore of their governing mainland countries, however, have a more difficult problem. Participation for them in their national governments is often logistically and politically difficult. Islands and their populations are too easily discounted nationally, with little or no chance of international participation.

The Islas de la Bahia (Bay Islands: Utila, Roatan and Guanaja) off the north coast of Honduras, form one of the eight Honduran government departments. Severely affected by hurricane Mitch in October 1998, it will be interesting to compare their rate of recovery with that of the other severely affected mainland departments (ReliefWeb Agence France-Presse, 1998).

Large disasters in large countries and in urban areas have nevertheless become established as the basis for public and official opinions and action concerning generalized response. It is these disasters that have become either the emotive persuader, on the one hand or, on the other hand, the equally persuasive medium for despair with regard to the apparent impossibility of doing anything about disasters of such destructive and disruptive power.

In either event the small disasters, which recur much more frequently than the large ones, and affect similarly large numbers of people in total, escape attention and escape international action. It is these disasters in the islands of the archipelagos, in remote mountain villages and in the fishing communities on the maritime coastlines that are losing out to the international global programmes. Islands could inform the continents, were they to be given the chance, by a reversal of this unfortunate global norm.

The major part of this book is based upon islands' experiences and experience of islands; and the five case-studies in Part II are all from islands or island states.

Climate change and sea-level rise

There is now less uncertainty than there was (Lewis, 1988b, 1994e) with regard to the acceptance of the phenomenon of climate change and one of its most significant consequences, that of sea level rise. Internationally accepted assumptions for increased mean climatic temperatures were established in 1988 as 1.5–4.5°C, and a sea level rise of 20–140 centimetres, before the end of the twenty-first century. These values are projected to rise steeply beyond this date to suggest sea levels much higher in the longer term. Interim accepted assumptions are for a 1.5°C temperature rise and 20 centimetres rise in sea level by the year 2025 (UNEP, 1988). Variations are now beginning to emerge from regional, as distinct from global, research, which have relevance for atolls and other islands, and for the major deltas of the Nile and in Bangladesh (Houghton, 1997).

Recognizing their extreme physical vulnerability to sea level rise, island states' governments met at the Small States Conference on Sea Level Rise in Male, the capital of the Maldive Islands, in 1989 (Lewis, 1990). Some island states are essentially single islands (e.g. Barbados and Sri Lanka); others are groups and archipelagos of several islands (e.g. Tuvalu), hundreds of islands (e.g. Tonga); or thousands of islands, as are the Maldives. Some islands and island groups are mountainous (e.g. Dominica) and some may contain active volcanoes (e.g. Savo in the Solomons, Niua Fo'ou and Kao in Tonga, and Soufriere on St Vincent or on Montserrat (Lewis, 1998). Many island groups comprise a variety of island types (e.g. the Cook Islands and Tonga). Others, for which sea level rise is especially threatening, consist entirely of atolls and reef islands (e.g. Kiribati; the Maldives; Tokelau and Tuvalu (Wells and Edwards, 1989; Lean, 1994).

Tuvalu comprises a chain of nine atoll islands, all but two of which surround a lagoon. Only one island encloses its lagoon entirely, the majority being made up of innumerable pieces of land (*motu*) surrounding their lagoon and each separate from the other as the atoll rim dips below sea level and reappears – in many places not much wider than single track roadway width. (One island has no lagoon but a swamp at its centre.) Distances across the islands' lagoons are 15–18 kilometres and distances between each island complex are 125–150 kilometres.

The entire atoll chain extends over 700 kilometres of ocean but with a total national land area of only 24 square kilometres divided between nine islands, and divided again many times, separating again these already tiny and fragile landforms and communities. The fragility of existence in a massive and powerful ocean is at once graphic and extraordinary (Lewis 1989a).

The largest single island is five square kilometres; the highest point of all islands is 4.5 metres above mean sea level – most land areas are appreciably lower.

The population of Tuvalu is 8500, 2700 of whom live on the principal atoll of Funafuti at a density of 1150 per square kilometre – equal to that of Malta. National population density per square kilometre is 347, one-third greater than that of the United Kingdom. A fragile economy, of which the only export income is from copra (35 000 Australian dollars in 1986), is stabilized by the Tuvalu Trust Fund (Australia, New Zealand, the United Kingdom, Japan, Korea and Tuvalu), income from the sale of postage stamps, and remittances from Tuvaluans overseas. 'Tuvalu's economy is small, fragmented and highly vulnerable to external influences': as is its economy, so is its topography and population.

If internationally accepted assumptions are correct, most of Tuvalu will be inundated by the end of the twenty-first century. As uncertainty will prevail for the greater part of 100 years, two points of view will ensue. One will argue that, with a sea already rising around fragile islands, continued investment in the development of a fragile existence now doomed, has no usefulness, and will induce people to stay in an increasingly hazardous environment. There are many Tuvaluans who would agree and who see the urgent need for the easing of immigration restrictions in New Zealand and Australia – 'where a few thousand more people would not be noticed'.

The other point of view will stress uncertainty itself and the 100 years in which that uncertainty could be prolonged; the realism of evacuation and Pacific precedents where many have preferred defiant and hazardous isolation to the unknowns of relocation (see Case-study I); and the experience of Tuvaluans of the sea and its hazards as a basis upon which to adjust over time. The imagery of possible ultimate catastrophe should not be made to preclude seemingly minor measures on behalf of the interim real condition.

The first effects of a rising sea level will not be new to Tuvalu. Construction of the Funafuti airstrip by American Forces in World War II destroyed the 'lens' of fresh water in the coral rock substrata. Ancient pits filled with vegetal mulch for the growing of root crops (*pulaka*) in otherwise infertile coral sand were the first to show the effects of consequent salination that has been worsening since, and which a rise in sea level will exacerbate. Efforts to introduce sweet potatoes, grown hydroponically in mounds of sand at ground level, will introduce alternative root crops – for the time being less vulnerable to rising sea water salination.

Neither is sea water flooding a new phenomenon; at the twice yearly high tides of February and September, parts of densely populated Funafuti atoll are flooded to depths of more than half a metre. Traditional house forms provided a floor level a metre off the ground – appropriate on land prone to flood. New 'Western' house styles introduced concrete floors at ground level and displaced outmoded traditional forms. New building codes have reintroduced floors raised significantly above ground level.

These innovations in food production and house building are being made with regard to indigenous and known hazards. Similarly, construction to prevent coastal erosion has been commenced – not on account of a rising sea but a normally aggressive one. Sea level rise will not at first present sets of hitherto unknown conditions, but intermittent exacerbation of these known hazards perhaps more frequently and perhaps more intensively but with periods of normality in between.

Funafuti atoll was overwhelmed in 1972 by the 50-foot waves of hurricane Bebe riding on an exceptional spring tide and accompanied by winds of up to 150 knots (Ball, 1973). Nearly all the 125 village houses were destroyed, and government buildings were damaged beyond repair. Five people died and 700 were made homeless; crops were annihilated and copra production fell by 80 per cent.

Vulnerability to hazards of this and lesser kind require a people able to cope and an infrastructure able to support them. The condition of both before catastrophe is significant as an enabler of survival and recovery afterwards. Provision of fresh rainwater for drinking and cooking; the removal of breeding places for mosquitoes and other vectors; rubbish disposal and general attention to environmental health are all factors of quality of life that become crucial to survival and recovery after any kind of disaster.

On Funafuti atoll, the population increased by about three times during reconstruction after the 1972 hurricane (and in anticipation of national independence in 1978). Around Vaiaku, the capital and seat of Government, there presently exist conditions of overcrowding, environmental degradation and consequent environmental health hazards, comparable in their aggregation more to some urban peripheries of major cities than to tropical atoll islands.

Tropical cyclones (i.e. hurricanes or typhoons) develop from interactions between sea surface and air temperatures. Sea surface temperatures of greater than the critical 26°C are expected to occur more frequently and over larger areas of ocean. An increase in the incidence of cyclones and normal storminess now appears to be more certain (Lewis, 1996; Gribbin and Gribbin, 1997; Maslin, 1998).

A strategy for development is required to accommodate a comprehensive approach to the implications of sea level rise in atolls and other islands. Social and cultural aspects that will render vulnerable communities better able to adjust to their hazardous environments, as well as the environmental and practical implications of sea defences, must be considered. When there is a national policy for development to take account of the early implications of sea level rise, a balanced long-term strategy can effectively commence. Matters of long-term consequence would best be considered when short-term threats have been accommodated.

Rather than distantly and dismissively to 'write these nations off the map' (Lean, 1988 quoting J C Pernetta), field examination of their context, condition, and culture may lead to more creative assessments of their future – certainly for the short and medium term, and possibly for the more uncertain long term as well.

3

The experience of vulnerability

VULNERABILITY IS EXPERIENCED in a range of contexts or at various 'levels' or sectors; national and sub-national (regional), social (district, community or domestic), institutional, military, systems and networks, economic and environmental. (Parker *et al.*, 1997 use similar headings). We also speak of supra-national regions being vulnerable, e.g. African drought or Caribbean hurricanes. Each level of this experience impinges on other levels; the national level upon the domestic, for example by failure of services and resource supplies; the domestic level upon the national, for example by dependency and inability to participate and contribute by work and production.

Dependency
At each level, susceptibility, and therefore vulnerability, is increased by dependency. That is, reliance upon external assistance or upon systems not of indigenous origin (see Case-study II). Dependency is a very real cause of vulnerability, whether it is by choice, chance or inevitability. The removal, disappearance or failure of the dependency 'prop' results in a worsening of a vulnerable condition; the converse of dependency being self-reliance and indigenous systems or cultural norms for the sharing of losses and of efforts for survival and recovery.

> Changes in resource management through time have led to growing dependency on external resources, both for food supplies and for a growing range of non-food products. This has also extended to include housing, where built of imported materials. Increasingly heavy incidence of hurricane damage has served to highlight these shifts in dependency. Whereas up to the 1960s it was possible for islanders to handle their own recovery, this was no longer possible from the 1970s onwards. Islanders found themselves desperately short of food, and relief supplies were required in large quantity and over a wider range of needs (Campbell, J.R., 1977, writing of Fiji).

Dependency theory was a forerunner of what is better understood as sustainability; one form of dependency is the reliance by this generation upon the resources of the next. In the longer term therefore, dependency can be a product of unsustainable development (see Chapter 7).

National vulnerability
National vulnerability is not only a matter of natural disasters, but the impact of natural disasters does have its place in concerns to do with threats to environmental security; other threats being those to territorial, political and economic security, for example (Commonwealth Secretariat, 1985, 1997).

Comparison of the impact of disasters at national level is an elementary step towards analyses of vulnerability to them. Small countries are physically most

vulnerable to their impacts, and poor countries have been shown to be economically most vulnerable – by comparisons of the estimated cost of disaster against
gross national product (Kates, 1980; Lewis, 1991b). The smallest poorest countries are the most vulnerable of all, a factor long recognized by those countries
themselves in the measures that some of them have taken (Lewis, 1976).

National vulnerability is a reflection of prevailing socio-economic conditions
in addition to location in relation to the incidence of hazards – just as it is at any
other level or sector of vulnerability. It is no coincidence that the Case-studies in
Part 2 emanate from islands and island states. The high proportional impact of
natural disasters upon island states is the subject of Case-study I (see also Lewis,
1991b).

Vulnerability may also prevail by sector – social; institutional; military; system; economic; environmental. Sectoral vulnerability is rarely independent of
other sectors and conversely, monosectoral perceptions of vulnerability usually
serve to reduce the range of perceived options (Parker et al., 1997).

Social vulnerability
The vulnerability of communities and people is usually manifest in social groups
that may be more vulnerable than others. These may be integral members of a
community, such as the very young or the very old, or distinctly separate groups
identifiable by settlement, ethnicity or religious differences, for example.

Being concerned particularly with the measurement of vulnerability in conditions of poverty in Andhra Pradesh, Winchester (1992) makes an in-depth
examination of social vulnerability which he considers arises from the sudden
'shocks' of external factors – such as illness, accidents, births and deaths, natural
disasters and civil disturbances – impinging upon already unstable conditions
accrued over time.

For the purposes of measurement, it is necessary for Winchester to distinguish
between poverty and vulnerability which, as he states, are inextricably linked.
Vulnerability comes not only from being poor, but from being powerless to do
anything about vulnerability which results from the actions and activities of
richer, and therefore more powerful, groups. The controls exercised by the rich
over the poor in their arrangements for share-cropping, water distribution and
money lending, for example, are evident examples of this process of vulnerability
accretion in South Asia. Similar examples abound in one form or another in all
societies.

Measurement of vulnerability is one thing; identification of its causes over
time is another. Measurement must be repetitive when what is being measured is
so relentless and so pervasive.

Institutional, military and systems vulnerability
Institutions with their own different and separately identifiable vulnerability are,
for example, military personnel, infrastructure and hardware to damage or loss
in use, and educational buildings to extremely hard wear in use. Political institutions (e.g. governments), however, have been known to fall, and/or coups and

assassinations to ensue, in the aftermath of, and as the indirect result of, natural disasters.

An early example of the political consequence of a hurricane is described by Robert Louis Stevenson (1892) in his description of the effect of a hurricane in 1889 on German, American and British ships assembled in Apia Harbour (Western Samoa) on the brink of war. All but one were wrecked and, according to Stevenson, the Treaty of Berlin resulted instead.

During World War II, a typhoon in 1944 crippled the United States Pacific Fleet in its support of the planned invasion of the Philippines. Three destroyers sank and 28 other ships were damaged; 146 planes were lost and a total of 790 men lost their lives (Brindze, 1973).

The November 1970 cyclone, and its subsequent alleged mismanagement, was one of the many influences that triggered the Bangladesh War of Independence which commenced in March 1971; and 'many a political reputation was made [in the aftermath of disasters and] during emergency relief operations' (Islam, 1992).

The earthquake that destroyed much of Managua in 1972 triggered in its aftermath the armed uprising, ensuing civil war, and eventual assassination of Nicaragua's President Samoza.

The currently prevailing risk of electronic systems failure at Year 2000 (Y2K) is the ultimate and imminent example. Billions of systems worldwide could fail simultaneously at the millenium. Those systems at risk range from domestic video cassette recorders and microwave ovens, to telephone exchanges, railways, hospitals, banks, oilrigs and satellites. The likely intensity of disruption is unknown, but it could be the cause of widespread chaos to supplies of electricity and food, and cause failures in transportation and financial networks (Huggins, 1998). All of these and more are at risk of widespread systems failure at the millenium – but they are all vulnerable from other causes at any time.

The millenium risk is that of failure within themselves of electronic systems; as distinct from failure brought about by external impacts – though failure within the electronic sector will impinge immediately upon all others which have un-avoidably become *dependent* upon it (for discussion of dependency and vulner-ability see Chapter 3).

Individual systems vulnerable in themselves may be in telephones and com-munications, computerized technology and networks, transportation, stockmar-ket trading, etc. The failure of one system (e.g. electric power; Park and Bender, 1990) may be the cause of failure of other systems; and vulnerability to these failures may impact upon social and institutional vulnerabilities (Wintzel, 1979).

Economic vulnerability

Economic vulnerability may be in agricultural production, especially in fragile monocrops such as bananas or sugar cane, common in many cyclone-prone countries. The proportional impact of damage in island states is invariably high; after hurricane Allen, damage as a share of GDP ranged from 15 to 80 per cent in affected island states (Lewis, 1991b and see Case-study III).

Changes in land use and cropping can show increased vulnerability over time. New emphasis on cash cropping can take over good quality land once given to food crops, which become forced to occupy lesser quality soil on more distant, marginal land more exposed to hurricane and erosion. Cash crops are of little use when food supplies are scarce (Bayliss-Smith, 1977).

Environmental vulnerability

Environments are vulnerable in themselves, and vulnerable environments have implications for the vulnerability of inhabitants and populations. Populations having become victims of their vulnerability impose their own environmental damage. Trees destroyed by hurricane may expose soils to erosion or may deprive a community of a food or monetary source. Excessive use of firewood for fuel may exacerbate desertification and over-intensive land-use to create conditions conducive to soil erosion.

The creation by the sea, suddenly or over time, of natural embankments, may create a protection that did not formerly exist. In Tuvalu, hurricane Bebe in 1972 caused a massive upsurge of coral debris to form a 19-kilometre embankment on Funafuti Atoll (and on other atolls) which increased the island land mass by a fifth and created a permanent embankment (Baines and McLean, 1976; Lewis, 1988b). The long-term behaviour of the embankment that protects Chiswell on the south coast of England, and its political, social and technological implications, are described in Case-study V.

All of these sectors and elements interact in their vulnerability and in that of the populations upon which they impinge and which impinge upon them. The connection of institutions and systems with political power suggests that the vulnerability of these does transfer 'down the line' to lesser institutions, and to communities and families that are in a variety of ways dependent upon them and upon the services they do or do not provide.

4

The making of vulnerability

VULNERABILITY CAN BE created inadvertently by development and change which may displace communities, cause planned or spontaneous relocation of settlements, expose to familiar or unfamiliar hazards, destroy or remove natural protective features, or reduce access to traditional resources.

Population displacement

The abstraction of figures of internal displacement caused by development projects, from displacement caused by other pressures, is difficult. The magnitude of development-related displacement has, however, been shown (globally, however) to be enormous (Cernea, 1996a). In only two development sectors, those of dams and transportation infrastructure, the world-wide total of displaced populations from these causes over the last decade comes to at least 80–90 million people.

In China over three decades (1959–89), the number of people displaced by water conservation projects (alone) exceeded 10 million. In India, during the past four decades, 18.5 million people have been displaced by projects for dams, mines and industries, and wild life reserves (Fernandez, quoted in Cernea, 1996a). Seventy-five per cent of those people had not been 'rehabilitated'.

'Impoverishment' results from these displacements, comprising the results of eight typical processes of landlessness, homelessness, joblessness, marginalization, loss of access to common resources, food insecurity, increased mortality and morbidity, and social disintegration (Cernea, 1996a, 1996b). Impoverishment is a process of deprivation and denial. The result of impoverishment is vulnerability to whatever happens next; the processes of impoverishment are the processes of vulnerability. Other examples of processes which may cause or exacerbate a vulnerable condition are also common in commercial agriculture and coastal tourism.

Commercial agriculture

Field research undertaken in 1980 (Jeffery, 1981b) examined the long-term history of the accretion of socio-economic vulnerability in Martinique and the Dominican Republic. In Martinique, a greater vulnerability to tropical cyclones and storms began for the indigenous population after French colonists defeated the Carib Indians in 1635 and drove them to the Atlantic side of the island. Mid-seventeenth century maps show a formal division of the island with a *terre des Français* to the west and a *terre des sauvages* to the east. Later, all Indians were removed from the island and slaves were imported to take their place. Greater exposure and vulnerability to storms continue in eastern Martinique for certain low-status groups.

The Dominican Republic occupies the eastern two-thirds of the island of Hispaniola; Haiti occupies the western third. Crucial changes that increased the vulnerability of many rural populations occurred during the era of President Trujillo, who exercised complete and personal control over the Republic from 1930 to 1961. In the aftermath of hurricane David, fieldwork was undertaken in the communities of Monte Plata, Bayaguana and Sabana Grande de Boya, all in the Province of San Cristóbal. These areas were selected because they were not the worst affected by the hurricane, where socio-economic indicators of differential vulnerability may have been obliterated.

Much of Monte Plata is low lying and suffers severe flooding in hurricanes. The causes of flooding are not a simple matter of topography, but are due also to deforestation that has taken place in this century to clear land for the mono-cultivation of sugar cane. Although sugar companies had been active for 60 years, widespread land clearance did not take place until the 1940s and 1950s during the Trujillo era. (Additional examples of innexorable marginality creating and exacerbating vulnerability in Egypt, Senegal, Sudan and African Sahel are given in abstracts from Mensching, Geist and Kishk in IDNDR, 1998.)

Erosion and flooding

La Caguaza, for example, in the low-lying flood-prone areas next to the Ozama River, had been a community of small-scale cultivators, with production focused on subsistence crops combined with cocoa, coffee and bananas, and the raising of pigs and cattle. In 1957, Trujillo sent in bulldozers without warning and all existing crops were destroyed, together with extensive woodland. More than half of the 22 households were dispossessed and obliged to leave. No compensation was paid, although the land was said to have belonged to those who farmed it.

This area, and others like it, are today entirely treeless and sown mostly with sugar cane owned by the State. The river banks, now unreinforced by tree roots, are more prone to recurrent erosion and are consequently much lower than they were before deforestation. As a result, the rivers burst their banks more frequently and not only as the result of hurricane rainfall.

Nowadays, communities such as these are no longer self-sufficient in food-crops. Although in recent years some land has been returned, the best land has been retained for sugar cane. Some land now used for food crops is that which is prone to flooding and therefore not suitable for sugar cane. This is now marginal land, not only in a physical sense because it is on the littoral edges, but also economically, because it cannot be relied upon to produce harvests, which may be destroyed in flooding. Basic food supplies have now to be brought in from outside the area, causing these communities to be dependent upon external sources – which may at times themselves be unreliable.

This is the marginalization process in action. First the land was taken away from diversified farming; then it was deforested, causing increased proneness to erosion, which in turn caused flooding to be more frequent; this increased the vulnerability of food crops grown since the land has become available again, causing the community to be dependent on outside sources and no longer self-reliant, as it once was.

Those who were removed, or who migrated from the area, were obliged in many cases to move out of the level valleys that had been taken over for sugar cane and to migrate either to urban areas or to previously uncultivated hillsides. This process, which has been the cause of over-farming and erosion of hillsides and consequent silting of river beds, accelerated after Trujillo's assassination in 1961 in the comparative 'free-for-all' that ensued. This also led to more deforestation, land clearance and hillside occupation.

Urban shanty settlements

Dispossession of small farmers was caused more by physical force than by market forces, even though land prices rose due to the cash-crop predominance of sugar cane and food prices rose due to scarcity and dependency upon them. Large numbers migrated to urban areas in search of work.

In Santo Domingo, the state capital, there are two principal areas of shanty settlement; one on the low-lying land along the Ozama river and the other on steep ravines on the northern edge of the city.

At times of hurricane, these shanty settlements are the cause of considerable problems for the government, since inhabitants have to be evacuated each time so as to avoid heavy loss of life. In 1980, torrential rains threatened to wash houses into the gullies, but there was a noticeable lack of comprehension as to how and why these communities came to be where they are. The administration that deals with the shanty dwellers on these occasions, does so only at the level of cyclone preparedness, and does so because that is the limit of its capabilities as well as of its comprehension. As a branch of government, it is not part of a developmental sector that might recognize the causes of the growth of migration to shanty towns, as well as having the powers and resources to do anything about it by way of relocation and/or rehousing and the creation of employment opportunities.

There are similar 'shanty towns' adjacent to thousands of cities, on ravines, hillsides and dry river beds, and all of them are known to be highly vulnerable. What is important, however, is for the reasons for their occupation to be known and understood so that governments may undertake measures other than evacuation and humanitarian assistance.

Understanding of these processes reveals people to be not only the victims of hurricanes and floods, but also as social groups and populations with a potential that, in the past, has been removed or restricted. Restrictions have the effect of limiting the capacity of a population for activities in support of its own self-reliance, and of limiting resources and power to make and to implement decisions. Though they may often appear to be within the remit and therefore to be the fault of those who have been restricted, most restrictions of this kind will have been imposed for the socio-economic betterment and protection of others.

Coastal tourism

Coastal development for tourism is common in many countries prone to tropical cyclones and tsunamis. Such developments are automatically at risk of flooding by sea surge as a result of tropical cyclone or earthquake (Lewis, 1984c). In many cases, however, construction activity destroys protective mangroves and trees by clearance, and dunes and reefs by quarrying and excavation for construction materials. Not only is the development itself at risk by its choice of location, but its own physical vulnerability and that of adjacent communities is increased by removal of natural protective features. Bearing in mind that development of most kinds draws an increase of population for the employment prospects it offers, vulnerability is at once compounded.

A project for the rational and ecological use of land resources in Fiji (UNESCO, 1977) commented:

> Nothing has been done to discourage extensive tourist development close to the shore. . . Use of shoreline sites is on the increase in Fiji, putting investments, visitors and Fijian people all at peril from rare but quite possible events.*

The currently (1998) proposed US$4.4 billion development, largely for tourism, of the Nungwi Peninsula of Zanzibar, off Tanzania, is not currently an area prone to tropical cyclones. Sea surface temperatures do however already reach 26°C at certain times of the year and could do so over increasingly wider areas of the east African coastline as a result of climate change (see Chapter 2) beyond the relatively small area north of Madagascar currently affected.

The development itself, with its visitors and staff, will be at risk, but more particularly so will be the thousands of displaced former occupants of the peninsula.

Disasters as agents of future vulnerability

Damage caused by one disaster, either regional or local, will in all likelihood render affected communities more vulnerable to recurrence of disaster – of the same or of different kinds. Similarly, response after one kind of disaster can very easily add to or create vulnerability to another kind. Fire was once the recurrent and prevailing hazard in St Johns, Antigua, where buildings were constructed of timber with shingle roofs. As a result, those who could afford to adopt masonry construction did so for stores, churches and private dwellings. When in 1843, earthquake occurred, it was the masonry buildings which suffered most, the flexible timber buildings being more resistant to earthquake movement (Lewis, 1984a; see Case Study III).

In the Cape Verde Islands, small dams and water conduits constructed in response to perpetual drought are among the first to sustain damage by earthquake and tremor, exacerbating an already fragile survival viability (Lewis, 1982c).

---

* The Fijian Government recently announced financial concessions to investors interested in building hotels in the country. Fiji's Prime Minister, Sitiveni L Rabuka, said that Fiji intended to increase revenue from tourism by attracting investors willing to construct facilities catering for tourists (*World Architecture*, No 50, October 1996).

Squatter settlements are invariably vulnerable to flash-flooding, wind and landslide, as well as earthquake, due to their occupation of the only accessible land in ravines, dry river beds, steep slopes, coastal and river margins, otherwise unowned or unused. Earlier disasters may have been the trigger for migration to city squatter areas. Among those affected by the 1977 flooding in Karachi were former Pathan hill dwellers who had migrated after earthquake and landslide in 1974; easily identifiable in the formerly dry river beds they had occupied, by their stone-built dwellings (Tapner, 1977). Where processes of this kind occur, only to persuade development to build more strongly against earthquake or high wind or to institute warning systems, can be a dangerously incomplete 'solution' which will suggest a safety that does not in fact exist.

The interrelationships between damaging events are as significant to the assessment of vulnerability as is the nature and likelihood of each one. Vulnerability assessments therefore are a necessary combination of research in scientific, technological, and sociological data, where available. Vulnerability is a morphological, cumulative and collective condition. It is not one which results only from each isolated damaging possibility as it becomes apparent or manifest (Lewis, 1986).

Historical analysis further exposes the interrelationships of one set of conditions with another in the making of vulnerability. Recently emancipated slaves and their new self-built settlements at St John's were identifiable as being most particularly affected by the Antigua earthquake of 1843. Similarly, it was not until after repayment (to the colonial government in London) of loans necessary for reconstruction after the earthquake, that construction could start of a reservoir to provide drinking water for St Johns in times of drought (Lewis 1984a and see Case-study III).

Hurricane Isaac in March 1982 destroyed 22 per cent of housing stock throughout the Tongan archipelago. A project identification mission (Lewis, 1983b) found that in spite of an impressive rehousing programme, more than half of those made homeless had not participated, being unable to afford their contribution of a quarter of the cost of a dwelling. Only at that stage was it visibly evident that a major long-term housing problem pertained, which established practice could not solve; and only at that stage had the problem become locally recognizable and a significant concern. Accurate information on numbers of houses destroyed and numbers of families accepted on to the rehousing programme was only then becoming available, prepared at the request of the mission.

The policy of requiring financial participation by the homeless in the reconstruction programme, though perhaps reasonable from certain points of view, caused the many who could not afford to participate to remain the most vulnerable to subsequent cyclones (Lewis, 1989b).

The same post-hurricane identification mission coincided with a period of severe water shortage which affected the entire archipelago, but which was most severe in the islands of the Ha'apai sub-group. Food crops, replanted in the aftermath of hurricane, were severely depleted due to drought. Fishing, the traditional standby when other food is in short supply, was impeded because boats and equipment had been destroyed or lost in the hurricane. In June 1983

31

fishing boats were still urgently needed for survival in drought. The mission was able to report on the local severity of the water shortage and on the need for boats and fishing equipment. It also recommended a project for hydrological investigations of catchment and ground-water availability and needs for current, projected and emergency requirements. The social, nutritional and environmental health implications of water supply were considered, as were the economic implications for copra production of greatly increased consumption of coconuts for their liquid during times of water scarcity (Lewis, 1983a).

Conflict and vulnerability

Vulnerability to the socio-economic consequences of conflict is closely comparable to vulnerability to the effects of natural disasters. As with natural disasters, however, conflict and war can also be creators of vulnerability to other hazards (see Chapter 7). The following examples are taken from past conflicts, although similar processes are currently and catastrophically self-evident.

Algeria
Algeria has a remarkable modern history of population movement. Keith Sutton (1969) and Richard Lawless (with Sutton, 1978) have given graphic and detailed analyses of the French policy of *regroupement* during the War of Independence (1954–1961). As part of a campaign against guerrilla groups, forced removal of rural populations, upon which the guerrillas relied for support, was initiated. Extensive *zones interdites* were created and forbidden to Algerians, who were shot on sight within them. At least two million people were 'regrouped' in this way with long-lasting consequences. Many never returned to their former communities, which in many cases were destroyed by French troops. At first no facilities for rehousing were provided for the expelled people, and serious overcrowding of existing villages outside the zones resulted. When *centres de regroupement* were eventually established they were inadequate. The Sersou–Ouarsenis border region is described as having been subjected to particularly severe military operations involving *regroupement*.

Only 50 kilometres away, the El Asnam region (now Chleff) which suffered the earthquake of 1980, cannot have been immune. El Asnam itself was then (as Orleansville) and still is, a military garrison.

Forests had been a resource for the rural population, but those areas may have been up to three-quarters destroyed by heavy napalm attack. When replanting was effected, it was with strict conservation controls and an important component of traditional rural economy was thus removed, or remained inaccessible.

Before the Algerian War of Independence, traditional rural economies had been seriously disrupted by the expropriation of fertile land in favour of the French *colons*, parallelled by rapid growth in the non-French (Muslim) population. At the advent of Independence in 1962, traditional economic and social structures, as well as actual communities, had been seriously eroded. At the same time, 90 per cent of almost one million *colons* returned to France, and about

32

200 000 *émigrés* returned from Tunisia and Morocco. Since then, emigration of workers to France has continued and the Algerian population itself has increased annually (World Bank, 1980).

Migration to urban centres also increased, and urban population grew by nearly 7 per cent between 1954 and 1966. The population of Algiers almost doubled in that time, and lesser towns and cities similarly increased (Descloitres *et al.*, 1973). Reconstruction after the 1954 earthquake ensured the inclusion of El Asnam (Chleff) among this urban population growth, which reached 105 000 in the *daira** by 1966, and increased by 48 per cent to 156 000 by 1977, 11 per cent greater than the population increase of the *wilaya** as a whole (ONRS, 1980).

After 1971 there was a programme of 'agrarian revolution' in Algeria (Sutton, 1978) involving the redistribution of land formerly expropriated by the French, and the creation of agricultural co-operatives. Redistributed land was often a considerable distance from the beneficiaries' homes, involving planned movement into new villages. Reflecting its agricultural importance, by 1976 there were eight of these villages established in El Asnam *wilaya*, against an average per *wilaya* of less than four.

Population relocation and migration caused a fragile social context in which the 1980 earthquake brought about further dislocation. In January 1981, on the periphery of the military controlled damage-zone of El Asnam city, very large numbers of people had quickly established marketing activities. Expectations of employment in reconstruction served as an attraction to easily unsettled, underemployed rural people. Tented villages for people from the city made homeless by the earthquake, were also on the periphery of the city area. Crowded, cold, wet and muddy living conditions, to some extent serviced with food, water, electricity and clinics, created conditions in the post-earthquake *villages de toile* which could have been perceived as better than some normal rural living conditions, and certainly better than provision that had been possible at that time for rural earthquake victims. In all, the earthquake itself and these post-earthqake conditions, created a pole of attraction at El Asnam (Chleff), as well as having triggered possibly permanent migration to other urban centres.

Papua New Guinea

The harbour at Rabaul is 14 kilometres wide and is formed within a volcanic *caldera* into which the sea has entered at one side. It is one of only three deep landlocked harbours in Papua New Guinea. The town of Rabaul is also within the *caldera* and is surrounded by other volcanoes – The Mother, South Daughter, North Daughter, Vulcan, and adjacent to Rabaul, Matupi (Lewis, 1994c).

Vulcan was an island formed by volcanic eruption in 1878. In 1937 Vulcan exploded, reshaping the island as a cone and joining it to the New Britain mainland. Matupi erupted at the same time, smothering Rabaul and its harbour with dust; 500 people were killed and the entire town was evacuated.

---

* *wilaya*: equivalent to a French *departement*, subdivided into *daira*, composed of *communes* incorporating village localities.

The Bismark Archipelago, extending from New Britain, was claimed as a protectorate by Germany in 1884. Rabaul was established in 1910 as the headquarters of the administration of German New Guinea. Taken by an Australian force in 1914 at the start of World War I, it was later administered by Australia under a mandate from the League of Nations as the Mandated Territory of New Guinea.

In 1939, as a result of the 1937 eruptions, the administration relocated to Lae on the northern mainland coast. During World War II, Rabaul was taken by Japanese forces in January 1942, after which the town was entirely destroyed by Allied bombardment.

After World War II, the capital of the new Papua and New Guinea was established at Port Moresby in former Papua to the south west. Due to commercial pressure favouring the deep port, the northern administration relocated back from Lae to Rabaul, which has since been entirely redeveloped. Papua New Guinea became self-governing in December 1973.

In 1937, the population of Rabaul was around 5000; now, the population is 30 000. The deep landlocked harbour, formed by the tectonic activity which is the source of hazard, but favoured for strategic advantage by both commercial and military interests, has significantly preconditioned the fortunes of Rabaul – from prosperity to total destruction and back again.

Ethiopia

Intermittent warfare of a different kind, between the Mursi and the Bodi tribes of south-western Ethiopia, creates a context of regular and expected external attacks. Measures taken to ensure the physical survival of people and cattle serve to make the economy of the Mursi more vulnerable to climatic uncertainty (Turton, 1992). Cattle are a form of insurance against crop failure; meat can be eaten when there is nothing else. Cattle, however, are the targets of neighbouring Bodi, and cattle raids result in the killing of Mursi tribesmen. That the raids are more successful when herds and herdsmen are dispersed as a result of water shortage and grazing due to drought, illustrates the close and intricate relationship of war and vulnerability.

It is not only the actual raids, but measures to cope with the threat of raids, which have a long-term effect on the Mursi community well-being and consequent increased vulnerability. The withdrawal of cattle from the best grazing, and their concentration into more easily protected herds, brings greater risks of the consequences of water shortage and is more environmentally damaging. The concentration of herds adjacent to settlements creates conditions for the trampling of food crops; the threat of attack also reduces agricultural productivity.

Warfare, however, is seen here not only as a means of securing the resources of others or of adjusting populations to resource scarcity. Warfare is also a means of establishing and maintaining separate political identities of neighbouring groups. Warfare here is a cause, not a consequence, of political identity. For the Mursi, choice does not exist between physical and political survival; the only way they know of saving lives is to save their way of life.

These three examples are indicative of countless experiences of warfare in various forms in many countries. That natural hazards are a feature of many of these three countries does not make them rare examples. Vulnerability to natural hazards has been and is being exacerbated by the direct and indirect effects of war and civil conflict. Nevertheless, as in the cases of military vulnerability, where the outcome of war has been influenced by natural hazards (Chapter 2), it seems that natural hazards are more likely to be exploited by the makers of warfare, rather than natural hazards being a reason for not engaging in warfare. Though natural hazards and consequent disaster aftermath may be the cause or exacerbator of political instability, whatever the relationships of war and natural hazards, increased vulnerability for civilian populations will invariably ensue.

5

Survival, vulnerability and development

Survival – a neglected issue

RECOVERY AFTER ANY kind of natural or other disaster depends upon the number of survivors, their capacity to continue to survive, and their condition before the catastrophe happened. The condition prevailing before a disaster, of a person, structure, community, or society, is of crucial significance to the degree of loss, damage or destruction sustained and to the capacity to recover afterwards (Haas *et al.*, 1977).

In the context of bombing in World War II, social survival was a significant issue (Lewis, 1987c), but it is strangely unusual to consider human survivability in contexts of disaster reduction strategies. This is probably due to the assumption of the availability and dominant profile of post-disaster assistance. Neglected or not, survival and survivability are the crucial issues. Strategies for their achievement could with advantage be incorporated more overtly into appropriate development. This would be humanitarianism within development, not as a component of other activities apart from development. It would also express a human right (Whitehouse, 1996).

Without survival and the restoration of well-being, recovery is either impossible or difficult to achieve. Only some aspects of the pervasive socio-economic condition of vulnerability are disaster specific; vulnerability to one thing is largely vulnerability to another. Vulnerability to the socio-economic effects of earthquake is the same vulnerability to the socio-economic effects of tropical cyclone; and vulnerability to natural disaster is, by the same token, vulnerability to the socio-economic effects of war and civil strife (Lewis, 1994b).

Basic needs for human survival and its continuation are food, cooking facilities, potable water, shelter, warmth, treatment of injuries, environmental health and welfare; together with communication and information on what has happened and where to go for assistance (see Case-study II; Kavaliku, 1974; Finquelievich, 1987; Lewis, 1987c). In addition to the primary impact of disasters and the physical and personal damage sustained, is the need that then ensues for human survival to continue, requiring an availability of, and accessibility to, the same basic resources.

Provision for survival is much the same as provision for vulnerability reduction; and provision for both purposes is common to all disaster 'types'. Vulnerability is not disaster specific; neither are basic needs for survival. Socio-economic survival in one instance is socio-economic survival in another.

Development planning policy has the responsibility to ensure the necessary availability of, and accessibility to, basic resources. The means and the system for doing so must be established before disaster, because in disaster aftermath it is too late – except on behalf of the victims of the next and recurrent events. Disaster impact is governed by the prevailing condition.

36

Survival does not have to be coerced; it is basic to life itself – instinctive and endemic. Availability and accessibility of, and information about, basic resources for survival have to be enabled and facilitated by sensitive and creative development administrations, just as they may be eroded by insensitive or over-centralized administrations. In other words, so basic is the need to survive that its facility requires enablement rather than provision or control.

The concept of survival relates directly to development. Provision through appropriate development strategy of basic resources, the need for which is so often exposed by disasters, will improve the quality of life before and between disasters. Their accessibility after disaster, as an expression of self-reliance, will reduce dependency upon post-disaster assistance which may or may not be forthcoming, facilitate indigenous coping systems, limit the extent and duration of disaster conditions, and in these human and social terms will serve to reduce disaster. Many small local measures should not be abandoned in the face of the image of massive catastrophe.

Post-disaster assistance and vulnerability

The assumption that post-disaster assistance will supply the requirements necessary for survival after disaster, goes so far as to deny the developmental input necessary for the improvement of pervasive conditions, economic and social, physical and cultural, that would serve to improve normal day-to-day living conditions and render the provision of post-disaster assistance less necessary.

Post-disaster assistance *assumes* the presence of survivors and in doing so, sidesteps the need to induce and to facilitate survival in the first place, or next time. Post-disaster phases of 'emergency', 'relief', 'rehabilitation', 'reconstruction' and 'recovery' suggest a sequence of response, both external and indigenous, to each disaster occurrence. The popular preconception of 'relief' is of 'flown in' goods and materials in an 'emergency', but rehabilitation and reconstruction are necessarily indigenous undertakings, even when these phases have become dependent upon external inputs. Some illustrations from the history of 'disaster relief' provide a useful perspective.

It had been the practice adopted by both French and British Colonial Governments to make both grants and loans in the aftermath of cataclysms and disasters (Lewis, 1982c: Annexe 8). The metropolitan government in France distinguished between its *colons* and the *indigènes*, frequently reducing the claims of the *colons* and often increasing those of the *indigènes*. Although the authorities in Paris were repeatedly exasperated by claims for additional assistance due to cataclysms, when after one event, the claims of the *indigènes* had not been included in the application for assistance by the local (French) administration, these were added by the Metropolitan Government. The Parisian authorities were more benign than was the local administration regarding assistance to the *indigènes*, whom the Parisian authorities considered were less likely to 'speculate' on assistance that might be forthcoming. In 1906, a mission for the assessment of

37

damage from two cyclones in the Comores Islands accepted without reduction the claims for the losses of the *indigènes* with the comment:

> Les indigènes n'ont pas cherché a spéculer sur le secours qui pourrait leur être alloué par la métropole, ils se sont bornes à indiquer avec sincérité le montant de leurs pertes.

In Antigua, the loan sanctioned by Parliament in London for reconstruction following the 1843 earthquake was the subject of hard negotiation between the colonial government in London and the island governor in St John's. Repayments were raised by additional local taxes and repayments which ended only in 1868 – after 25 years (see Case-study III). It was not until the earthquake loan had been repaid, that construction could commence of reservoirs for town water supply for survival in recurrent drought.

In the Cook Islands, it was missionary organizations that introduced the concept and practice of externally mobilized relief after natural disasters, but again there were differences of opinion between headquarters in London and individuals in the islands. A severe hurricane occurred in February 1841, and another later in the same year. A disaster relief consignment of clothes was sent from London for distribution to 'orphans and other cases of real distress'. The Reverend Charles Pitman wrote in acknowledgement of the consignment (December):

> . . . but after all, dear Sir, generally speaking, the giving system is a bad one. There are many, as long as you will give, they will not work, plant, or strive to obtain what is necessary. If the people could get a sure market for what they could grow, I have no doubt that they would plant so as to obtain what was needful for their comforts, and what more is wanted? . .

The hurricane of 1846 was also severe. This time, relief consignments of food in the form of rice and biscuits were despatched from London. Pitman politely protested again and at length against the 'almost useless expenditure' and arrival of the food 16 months after the hurricane, and after abundant harvests in the meantime:

> Rice is an article of food to which the people here are not at all accustomed, and the want of utensils for cooking it, will be a great difficulty, as scarcely a person in our whole settlement possesses such a thing as a pot or pan to boil it in, their own food not requiring such articles for the purpose. . .

Pitman went on to express a strong preference for tools for reconstruction which he said would be 'invaluable'. (London Missionary Society Archives: quoted in Lewis, 1982c: Annexe 13).

Tonga had not been a colony as such, having been an independently self-governing kingdom and a protectorate of Great Britain since 1900. Colonial reports were filed by the consul (the nearest high commissioner being in Fiji) in which the concept and practice of disaster relief first appeared in 1909, after a hurricane struck the island of Niua Fo'ou: '. . . the Government of Tonga sent in relief but it was not required to any great extent' (Westgate, 1975 quoted in Case-study II). There was no further reference to (or need of ?) relief until after the hurricane of 1961, though there had been many severe manifestations of hazards in the meantime – including the eruption on Niua Fo'ou in 1946 (see Case-study I) and in a history and prehistory in which hurricanes, earthquakes, volcanic eruptions and droughts were endemic.

The question is put in Case-study I: 'Is disaster as such, in these situations (Polynesia), a wholly Western concept, introduced by alien administrations from alien sources and adopted for practical and pragmatic advantages?' Whether humanitarian or political in its inception, its acceptance was cultural and pragmatic. The result in most cases was the introduction of a previously unknown dependency and the root of a new-found vulnerability.

Vulnerability and development

Survival is the crucial prerequisite for recovery and reconstruction. Reconstruction is opportunity to improve, applicable to physical, institutional and social structures for which a surviving population is the prime prerequisite (Lewis, 1987c). Physical and metaphorical reconstruction in all sectors is the opportunity to reshape human settlements so that in doing so, survival from disasters is more assured.

Reconstruction of this kind more appropriately becomes development, embracing both the improvement and the institutional strategy for its achievement, so that both the condition of vulnerability and the processes which have led to it, and which could continue to lead to it in the future, are modified. Development policy is responsible for the condition of people, infrastructure, community and society and can, as part of its overall product, facilitate survival. On the other hand, inadequately conceived development can create the conditions for the exacerbation of vulnerability.

Vulnerability is pervasive; 'vulnerability' has implications beyond disaster discourse. Vulnerability to the effects of war and conflict is vulnerability to the effects of natural disaster. Development to reduce vulnerability to one will reduce vulnerability to the others. Often, activities which are the reverse of the desired objective, serve to emphasize the value of that objective. An indicator of the significance of resources and services for survival – personal, community, organizational and political – is that in many cases of conflict (for example in Angola, Ethiopia, Mozambique, Nicaragua, Somalia, Sudan and Uganda), military objectives have been the destruction of food resources by aerial bombardment, burning, mining, killing of livestock, poisoning of water sources, sieges of market centres, targeting of feeding centres and interruption of relief supplies. Attacks upon health centres and their personnel are similarly indicative of the value of these and other infrastructure towards the survival of the populations they served.

Basic-needs development appropriate to vulnerability reduction and survival, serves also to increase the quality of life between disasters. Where conflict is caused or exacerbated by perceived inequalities between populations and regions, equitable basic-needs development may also be the commencement of a process that renders conflict and civil strife less likely.

The effectiveness, and cost-effectiveness, of measures to reduce vulnerability is thus far higher than of those measures planned to resist or to protect against each specific hazard. This is the more so when it is realized that the needs of survivors to continue to survive are the same needs as those for viable, equitable, self-fulfilling, healthy, productive and sustainable communities (Lewis, 1987a).

The condition of a person, structure, community, or society before any kind of disaster, has a significant bearing upon its capacity to recover after loss, damage or destruction has been sustained. By ensuring that specific needs of survivors are more likely to be accessible to assist their continued survival and rehabilitation, a general upgrading and enrichment of communities takes place. Communications, social and health services, food and water supplies, shelter, training, education and information, required as resources for disaster aftermath, are what society regards as products of general improvement and development. Policies that affect the provision of this social infrastructure, and the way in which society develops and changes, bear directly upon the degree of its vulnerability.

Development in earthquake-prone areas, for example, has to acknowledge earthquake hazard, but also hazards of all other natural and man-made kinds. Concern with only selected or perceived specific hazards results in specific protection from, or resistance to, each one largely in improved physical and management structures. Because, however, complete removal of all hazards cannot be assured, at the same time the degree of susceptibility to hazard, which is vulnerability, must be taken into account.

In his book entitled *Disasters and Development*, Fred Cuny (1983) observed that vulnerability reduction 'will have little impact unless it is conducted in concert with normal development activities'. Cuny's approach was, however, essentially post-disaster and focuses upon how what is done afterwards can either impede or assist development objectives. But it is those development objectives which must, in the first place, take an integral account of hazards and of disaster probability. By ignoring hazard potential or by the assumption that development of any kind will make disasters go away, disasters could actually be made to increase.

There are countless cases where 'development' has made matters worse. Development is not a utopian panacea to be subscribed to at every opportunity – least of all by the tuning of 'disaster relief' to the support of development that may be largely negative. Does the importation of new foods, reinforced by their provision as post-disaster assistance, weaken motivation for the home production of subsistence food crops, and thus increase vulnerability? Vulnerability of populations and their physical infrastructure is discernible and identifiable before, and regardless of, the event of disaster; and development can be made to reduce vulnerability by taking on board the prevailing potential of hazards and their contexts.

Whereas risk is a product of the hazard – that is, the probability of earthquake, cyclone, explosion, etc. – the vulnerability of a settlement, community, group,

40

installation or structure, is a product of human ecology which renders it susceptible or not to damage and loss. The degree to which any of these elements can accept, absorb, reduce or change their vulnerability, and thereby accommodate risk, is an expression of their socio-political and socio-economic integration. What disaster is to one place, person or group, may not be so much of a disaster to another. Drought in Antigua with its disastrous effects on sugar cane production, would not have been regarded as drought in the Cape Verde islands, subject and accustomed to a much harsher rainfall regime (Lewis, 1982c).

Disasters are an extreme extension of events which are normal; the degree to which, for a particular person, group or community, an event becomes a disaster, being set by prevailing political, social, cultural and economic conditions. Is it not these conditions that development is all about?

There are thus four sets of occurrences within which vulnerability may be perpetrated, or by the same token may be reduced:

- external events such as natural disasters and wars
- natural attrition – deterioration and aging
- changes in socio-economic policies and conditions
- the extent of access to resources and services

and the interactions and interrelationships of any or all of these.

The role of development planning, programming and implementation has therefore to be to:

- take account of hazardousness at local levels
- rehabilitate and reconstruct damage, displacement and disruption
- analyse vulnerability at local levels, both by post-disaster assessments and by socio-economic assessment between disasters
- identify causes and processes, recognizing multi-hazard contexts that have created or exacerbated vulnerability
- devise and implement programmes and projects for equitable socio-economic vulnerability reduction.

These measures have simultaneously to be incorporated into development programmes. They also have to be interpreted as physical and social and economic measures. Projects for physical protection against exposure to the forces of natural disasters (e.g. shelters and embankments) and for resistant building construction, have to be regarded as components of programmes for disaster reduction – but not as their absolute entirety.

The reduction of vulnerability as the crucial path to disaster reduction has more to do with access to social and material resources, and with social participation linked to cultural expression and traditional knowledge and norms, than with 'protection' by technology in building construction, or in warnings and communications. Disaster prevention must be made inclusive of the enablement of human ecological adjustments in the activities of vulnerable people to maintain a resilience and self-reliance to counter the effects of disaster, rather than only as technological resistance to the forces of environmental extremes themselves.

Disastrous extremes are extensions of a prevailing normal hazardousness. Small and local disasters cause conditions of vulnerability which contribute to subsequent larger ones. *Ad hoc* response only to extreme and rare disasters is fallacious; it is the recurrence of small-scale frequent events that requires most attention, in themselves as local experiences and as creators of vulnerability to the large ones. Furthermore, the more frequent are events, the more normal they are – to the point where normality itself is the vulnerable condition. Thus, it is the more normal environmental condition, the enablement of survival in that condition, and the development of socio-economic contexts at local levels that are required, not only a globally compared and motivated response to the more unusual, emotive or spectacular disasters (Lewis, 1987a).

So much 'development' in response to natural disasters serves a simplistic objective of putting things back as they were before, or of simply providing secure places for people to go to. Neither of these objectives serves to change anything more than to save lives – a crucial consideration. The quality of lives saved, however, and the political, social and economic contexts of those lives, the poverty, destitution, oppression and ignorance prevailing upon and within them, do not change by these measures. Indeed, there are some who have observed that such measures on their own serve to exacerbate those socio-economic conditions which are the root of the vulnerability they aimed to reduce. In addition to the saving of lives, livelihoods must also be preserved and the quality of life improved (Lewis, 1997).

Disaster definition depends upon an event that renders a population incapable of recovery without external assistance. If, therefore, by a kind of development appropriate to reduction of compound vulnerability, it becomes possible to enable communities to better survive and to recover without external assistance, then disasters will have been reduced.

PART TWO
STUDIES OF VULNERABILITY

In a small place, people cultivate small events. The small event is isolated, blown up, turned over and over, and then absorbed into the everyday, so that at any moment it can and will roll off the inhabitants of the small place's tongues. For the people in a small place, every event is a domestic event; the people in a small place cannot see themselves in a larger picture, they cannot see that they might be part of a chain of something, anything. . . and they live like that, until eventually they absorb the event and it becomes a part of them, a part of who and what they really are, and they are complete in that way until another event comes along and the process begins again.

A Small Place Jamaica Kincaid. Virago Press. 1988.

Vulnerability and the analysis of context

THE CHAPTERS OF Part 1 have described how large disasters continue to attract the most attention. As each disaster occurs, its characteristics and its effects are assessed and compared. Globalized comparisons between singly assessed disasters are a reflection of a privileged view afforded to international but distant outsiders to each event.

The experiences of those directly affected, the insiders, and the nature and characteristics of the affected place which they occupy, appear as secondary in these global comparisons, whereas these characteristics are essentially and necessarily to do with local disaster impact upon them and their place – regardless of its disaster significance or otherwise to outsiders. This continuing state of affairs is not assisted by the reporting of disasters in the media.

The reporting of disasters

Disasters make news. Floods in India, China or Bangladesh; earthquakes in Japan or Iran; and cyclones in the Philippines, Bangladesh or the Caribbean may occupy prominent positions in national newspapers for several days.

Suddenness, and disasters affecting national and regional capitals, are likely to command headline attention. Other disasters, of similar impact but in less topical or 'newsworthy' countries, such as Nepal or Laos, may receive only brief accounts, not commensurate with their magnitude. By a straight comparison of their coverage, news items are not consistent or logical in the degree of significance, or otherwise, given to disasters. Two hundred thousand homes submerged in Vietnam would have rendered over a million people homeless and are 'worth' more than eight lines, when two million homeless in India get significant double and triple column coverage for ten days. This kind of comparison can be made in any selection of Western national newspapers at almost any time. Big disasters are news, but not all big disasters have similar news value.

Many major disasters are not given significant news coverage, but many minor ones are. Thirty-four people dead in a Texan tornado commands news attention, when thousands homeless elsewhere may not. News stories of individual survivors, saved against all odds, are the lifeblood of newspapers. Claribel Lovelace in the USA, was trapped for two hours in a six inch airspace, and Laura Arriola from Honduras was lifted from the sea, six days after hurricane Mitch had killed 11 000 other people. How many Claribel Lovelaces and Laura Arriolas were there in the Bangladesh floods?

Haphazard reporting of major events only occasionally succeeds in emotionally conveying what it is like to be a victim and inside the situation – being the aftermath. Being inside the situation is surely much the same, whether the event one is in receives headline treatment for ten days, or gets eight lines on only one

45

day. Even television, in moving colour, only occasionally makes the break-through from interested observer on the outside to involved participant on the inside. At any event, it is we who are the outsiders and they who are the insiders; it is the outsiders who do nearly all the reporting and talking. 'We are informed of everything but we know nothing' (Lewis, 1979a).

Newspapers local to the incident in question assist a conversion of outsider reporting to insider experience. Though invariably from outside, they do more to place the event in its local context. An analysis of press reports from Indian newspapers showed the comparative inadequacy (but not inaccuracy) of report-ing of the same event in Western newspapers.

Ratios of impact can also be made, either locally, regionally, or nationally according to availability of base data on population or housing, for example. The national, regional or local impact of a disaster can then begin to be demon-strated – not simply the world impact on news media. Laos or Vietnam are very much smaller countries than India; the national impact of one million people rendered suddenly homeless in the former is far greater than the impact of two million people rendered homeless in India.

Comparisons of country size are another crude indicator of national disaster impact. India is 1 260 000 square miles and Vietnam is 127 000 square miles. The impact of disaster, of similar or comparable geographical extent, on the small country will be far greater than the impact on the large country. Nevertheless, it is rare for news media to venture into such comparisons in their disaster report-ing, which remains inconsistent, though not unreliable.

There would be less reason for concern if inconsistency was contained within the realms of news reportage, but it is in the nature of any news that it becomes an influence on public opinion. News reporting from all media sources will become the basis for public response in contributions to relief and post-disaster assistance by non-governmental organizations that will appeal in the same media. No comparisons of national impact and no assessment of the stricken country's own national capacity to respond to its own disaster will be made – simply illustrations of the huge numerical impact of what has happened.

Prevailing conditions and characteristics

Part 1 has also shown how prevailing characteristics and social, economic and physical conditions are a significant contributory factor to natural disasters. Although such factors are invariably exposed to some degree by each disaster as it occurs, their identification, interrelationship and assessment, before and between disasters as analyses of contexts, is necessarily a significant part of the analysis of vulnerability.

The adoption of this approach facilitates a shift of emphasis from the place as a distant, transient and 'unknown' post-disaster phenomenon, to insider experi-ence and knowledge of the prevailing vulnerable nature of places and popula-tions, before and between, as well as after disasters have occurred.

Prevailing conditions will reflect the effects and experience of natural hazards over time, and of all phenomena. 'The hazardousness of a place' relates not to one phenomenon or to another, but to all hazards to which the place is prone. Within and as a part of indigenous understanding, the political, social, economic and physical characteristics of the place can be made to begin to adjust to comprehensive hazardousness as a matter of selective and specific strategy. The place takes on board its hazardousness and hazardousness becomes a part of its expression.

The case-studies that constitute Part 2 are an attempt to overcome the anonymity of small and distant places, to show how they can be investigated to expose to some extent the contexts upon which hazards impinge, and how the results of investigation reveal responses to those hazards and to overall and interrelated hazardousness. They commence with an account of the experience of one disaster in one small and remote place and continue with analytical perspectives of disasters in other small places.

These analyses of places, as the contexts of the manifestation of natural hazards, illustrate how contextual vulnerability analysis can be applied for the purposes of disaster reduction.

Island places

It is the characteristic of islands that their comparitively small area and extent is physically limited and defined by a natural containment. They are physically and administratively finite, though island states, as distinct from islands themselves, may be dispersed and complex, in contrast to the relative simplicity within each of their island parts.

It is in the nature of islands that they are a containment of interrelationships between activities. In that containment, connectivity between policies, activities and environments are more readily evident and more easily identified. Conversely, in the archipelagos, there is inevitable division and separation of activities that create special problems. Overall, the combination of interrelatedness and geographical division epitomizes the problems of government that all countries experience.

The advantage of smallness and containment within islands to contextual analysis, is that interrelationships between one event, action or activity and another, are evident and relatively unobscured.

Within these island entities, history has moulded, and has been moulded by, the effects of natural hazards. More significantly, the history of previous disaster occurrence is a key to the future. Rather than each disastrous event being recorded as it occurs and observed retrospectively, for each and all to be seen in interrelated social, economic and political contexts assures an opportunity for appropriate long-term understanding and adjustment.

Historical analysis facilitates post-evaluation of measures and consequences and of their interrelationship, their efficacy or otherwise, and the problems exposed or brought about by the swings of concern from one disaster aftermath to another.

That most island states are also less-developed (and some the least-developed) countries, confers an appropriate significance to development as the priority medium for change, and consequently, upon what the objectives of development should be to achieve predominant strategies for disaster reduction.

Islands are not always states in their own right; many island populations are components of larger island groups or of continental land masses. Such islands may either be endowed with national strategic significance or – politically as well as geographically – they are peripheralized. In either case, their expression of local vulnerabilities is inhibited. On Chiswell, off the south coast of England, since Roman times restricted access to the sea afforded access by the sea into habitation. Subsequent technological and military developments have eroded that delicate balance, and vulnerability to sea flooding has been exacerbated.

Understanding of context and its history affords a backdrop against which both short-term responses and new developments can be evaluated as parts of the crucial long-term morphology of a hazardous environment – inclusive of those developments misunderstood as not having anything to do with hazards or vulnerability.

Such characteristics and their evolution, appropriately identified in the analyses of small places, are relevant to disasters in places anywhere; large disasters are a fusion of myriad simultaneous small ones.

Recurring themes

The incumbent and the creator of vulnerability

The overriding issue to emerge from these case-studies, is that vulnerable conditions are rarely caused by the vulnerable incumbent. Vulnerability is more usually caused by the decisions, policies and activities of others.

Even where vulnerable conditions are closely associated with poverty, rarely is poverty itself caused by the poverty-stricken incumbent. Poverty is more often caused by the decisions, policies and activities of others – governments, institutions, commerce, and others either with, or seeking, power over others.

This means, therefore, that those whose actions lead to the creation of vulnerability have the potential and the power to take decisions and actions to reduce vulnerability. The principal process by which vulnerability reduction can be achieved is development, and ways in which development can be moulded to achieve that objective are discussed in Part 3.

Massive proportional impacts and small-scale responses

Even the 11 000 000 people made homeless in Bangladesh by the cyclone of 1970 cannot closely compare, at 15 per cent of the national population then, with the aftermath of hurricane Bebe of 1972, which made 22 per cent of the population homeless in Fiji – having a far greater national impact and causing far greater national suffering. 'Small' disasters are major cataclysmic events when experienced from within by the insiders at whatever level – community, regional, or national. (A report of hurricane 'Bebe' appeared in four column inches in *The Times* giving the first underestimated accounts of damage and numbers of homeless. In the same five lines, severe damage on Funafuti (Tuvalu) from the same hurricane was also reported as seen from a passing plane.)

Consideration of proportional impact confers a significance of disaster upon small places. By considering small places in this way, they come to be included in the global spectrum of disasters, from where they have so often been excluded in the past.

This principle commences the crucial process of transferring disaster assessment away from remote global comparative accounting, into the place where disaster has been experienced. Part of that experience, and part of the expression of that experience, is analysis of the impact that has occurred in relation to what existed before. For example, so many houses destroyed *out of so many that there had been*; so many people made homeless *out of a total population*. These data can then be expressed more significantly as a percentage. Only then, if international comparisons are needed at all, can they be realistic.

Assessments of proportional impact do not stop at national levels; they can also be made within sub-national and local levels, in administrative districts and areas. In Sri Lanka, by this process, some surprises ensued. It was not the coastal urban area in the full force of cyclone as it came in from the sea, but rural areas inland of it that suffered the greatest proportional (*and* magnitudinal) impacts of damage. Rural impacts are often as great, if not greater than urban damage; it is urban damage however, that is more impressive and more accessible. As in national comparisons of proportional impact, it was the smaller administrative units that suffered the most.

Among similarly vulnerable administrative units, allocations should be made commensurate with their populations, rather than the more favourable allocations being made simply on the basis of size to the larger ones – likely to be those of greatest political significance. When disaster does occur, it is likely to be the small units that suffer the greatest proportional impact and have the greatest problems. Exactly the same situation prevails at local levels as it does nationally; small units are likely to lose out unless proportional comparisons are taken into account.

This then leads to the scale of development projects. If projects are to be designed and programmed so as to be commensurate with small countries and small administrative units, then they too will need to be small – and the smaller they are, the more of them there will need to be. This will not go well with administrators who seek economies of scale. Although it may be more economic from the provider's viewpoint to go for large projects with minimum duplication of personnel and effort, from the viewpoint of the recipient it is preferable to achieve local integration by the widespread repetition of small inputs.

The need for the provision of basic resources on small and remote islands is stressed in Case-study I. The presence, at the time of an eruption, of a secondary school and a clinic, as well as improved communications and transportation facilities, would have increased resources available in emergencies at the same time as reducing the perceived need for long-term migration to the capital for access to these and other facilities.

Provision of communications facilities is required between remote communities, as well as between those remote communities and their capitals. In the South Pacific, the facilitation of traditional mutual assistance and support between communities, could be regenerated by the provision of boats and jetties, and do much to relieve dependence upon emergency assistance from the capitals or international sources. The intercommunication thus afforded would be of benefit to trade and exchange and to normal quality of life, as well as in times of emergency.

It is relevant to observe in this connection that, in the case of the two neighbouring archipelagos of Fiji and Tonga, some islands of each are closer to each other than they are to their own national capitals.

Recurrence

The scale of large disasters, perceived in their global context, tends to obscure the fact that, indigenously to their place, they recur. They may recur frequently or intermittently, on a large scale or small scale, seriously or less seriously. Globally, we are informed of a damaging earthquake here or a catastrophic cyclone there – 'not another disaster!'; but we are distracted from the realization that large disasters of all kinds are not only recurring globally, but that disasters both large and small are recurring in the same places, in some of which the effects of disasters of all kinds and sizes intermix and interrelate.

Concern for the recurrence of disasters applies equally to large and to small places, but in small places they are more likely to recur on top of one another – in the same places and not merely in the same country. The recurrence of, for example, cyclones in the Ha'apai sub-group of islands of the Tonga archipelago is both catastrophic and impressive.

The issue of recurrence itself recurs in all the case-studies, with the exception of Sri Lanka (Case-study V). Sri Lanka was selected for field study after the occurrence of one cyclone precisely because there had not been recent history of cyclone occurrence! Coconut palms were especially vulnerable, and caused damage to buildings as they fell because the trees had not suffered cyclonic winds before. They had not been periodically 'culled' – as trees had not before the hurricane-force windstorm that affected southern England in 1987.

Recalling that socio-economic vulnerability is as significant as the physical recurrence of kinds of hazard, all hazards are equally significant in their impact upon that vulnerable condition. Recurrence is therefore of generic disasters, not only of each 'type'; recurrence is the manifestation of all and any hazard. More accurately, it is a prevailing and continuous condition of exposure as normal hazardousness.

It therefore follows that what is done after one disaster will have a bearing on prevailing conditions for the next. More usefully, this can be interpreted to mean that whatever is done after one disaster should have in mind its role to reduce vulnerability to each and all subsequent disasters.

Interrelationships

Historical analysis allows a dissection of events and issues, but history serves a greater usefulness if it is also able to interrelate retrospectively in meaningful ways, the issues and events it has separated. This is particularly useful in examinations of the historic contexts of disasters. There are many lessons with which to be acquainted, such as the swing away from one method of construction in the aftermath of one disaster, only to increase vulnerability to another of a different kind; from timber to masonry after fire, and from masonry to timber after earthquake, for example. The importance of realizing the full extent of 'all hazards at a place' cannot be more real than as described from Antigua (Case-study III). Indigenous realization prevailed that drought could be as severe an occurrence as earthquake, but the construction of a reservoir had to wait for 25 years until the repayment to London of the loan for earthquake reconstruction.

Equally significant interrelationships of social and economic issues are evident in Tonga (Case-study II) and in Chiswell, Dorset (Case-study V). Not least, the interrelationship between emergency relief and self-reliance, and relief and development; the need to provide adequate drinking water catchment so as to preserve coconuts for the production of copra (Tonga); and between the policies and activities of one administrative sector and another, including those which appeared not to perceive natural hazards at all (Chiswell).

The importance of not making vulnerability worse also emerges from Chiswell (Case-study V). A perception of development for tourism on the one hand and the development of naval facilities on the other, imposed themselves upon community development, obstructing natural drainage of exacerbated sea flooding.

The crucial need for a balance between subsistence food crops and cash crops has been referred to (Chapter 4) concerning sugar cane production in the Dominican Republic. In Tonga, care has to be exercised with the production of vanilla, pyrethrum and passion fruit so as not to displace food crops, essential in emergencies if dependency upon imported foods is not to result. Throughout all the case-studies is the prevailing need for attention to political and social measures, as well as to economic, physical and technological measures.

Mutual protection

That buildings have the capacity to protect each other, is a recurrent feature from Sri Lanka (Case-study IV) and Chiswell (Case-study V). This factor has a bearing on the arrangement of buildings and on settlement planning, in the clustering or grouping of dwellings and other buildings and in their integration with other features such as new or existing trees. It also has implications regarding the retention of buildings, even when damaged to some degree because, when left in a secure condition, in storms and high winds they can continue to protect buildings adjacent to them.

Maintenance

The important, but to most people the tedious, issue of maintenance is most evident in buildings and other infrastructure (Case-study II). The wall that collapses in a storm and kills people sheltering against it, is a reflection not only of storm strength but of the condition of the wall. Although the casualties can conveniently be ascribed to the storm, they may more accurately relate to lack of maintenance of the wall. The funding of small works improvements becomes crucial in this respect.

Maintenance of buildings can ensure that less building damage will ensue and that consequently disasters will be reduced in this respect. That hazards also contribute as a cause of the need for maintenance simply emphasizes the requirement.

Maintenance is not only a physical concept to do with buildings. Maintenance is also required, for example, in agriculture, of irrigation and food storage, of water disposal systems, and in systems for communications and commerce. The concept of non-physical maintenance is probably best epitomized by the process of recurrent training programmes.

The case-studies

The case-studies which follow are based upon previously published articles. Each one has emanated from the undertaking of an assignment for advisory purposes. Each case-study therefore reflects an understanding from brief experience of a particular place, in relation to the natural hazards to which it is prone. The natural hazards referred to include tropical cyclones (hurricanes), storms and sea flooding, earthquakes, volcanic eruptions, fires and droughts.

The places in which the experience and impact of these events is described and analysed are Antigua in the Caribbean, Sri Lanka in the Indian Ocean, Tonga in the South Pacific, and Portland off the south coast of England.

All of these places are islands or archipelagos. This is partly to do with the selection of material, but significantly more to do with those places that requested advisory missions – usually in the wake and aftermath of natural disasters by which they had been afflicted. Requests for advisory missions were a direct expression of a need within small places – which were the first to realize the degree of proportional impact upon them.

The case-studies are based upon island places, and islands have the common characteristic of smallness. This does not imply, however, that what happens in islands is unique or special to them. Islands as small places are typical in many ways of small places anywhere, be they other islands or in continental countries. A theme of this book is that in the past there has been too much emphasis on disasters of large magnitude, and that those disasters for the most part have therefore been in continental countries. This emphasis upon islands as the places from which the essays derive, is a way of ensuring their relevance as small places to 'small' disasters anywhere and everywhere. Measures for disaster reduction that derive from small places are similarly widely appropriate.

The case-studies are in chronological order of their writing. They progress from an eye-witness account of one singular disaster on one small place, a volcanic eruption of a Tongan island (Case-study I), through historical analyses of natural disasters and of responses to them (II and III), field analysis of impact (IV), to an analysis of a sequence of storms and their effects in relation to seemingly unrelated undertakings and policies (V). All the case-studies, and the experiences they contain, have contributed to the approach to vulnerability as it is described in Part 1, as well as to the recommendations for development described in Part 3.

(Case-study references: references specific to the case-studies appear at the end of each one; general references are included in those at the end of the book.)

CASE-STUDY I

Volcano in Tonga*

THIS ACCOUNT OF volcanic eruption and subsequent evacuation in Tonga was given by Town Officer, Moeake Takai, who kept a diary of events at the time, and to whom the author is deeply indebted. Acknowledgement is made of the permission granted by the Ministry of Overseas Development (now the Department for International Development) and the Government of the Kingdom of Tonga for the original publication of this article. The experiences of the inhabitants of Niua Fo'ou during the eruption, the subsequent evacuation, and of the return of some of them, are recorded in detail in Rogers (1986).

The Kingdom of Tonga consists of 172 islands in the South Pacific, approximately 1000 miles north-east of New Zealand. The island group extends over an area of 140 000 square miles, but with land area of 289 square miles (see map of Tonga, Case-study II, Fig. 1). Only 36 islands are permanently inhabited, with a total population now of approximately 100 000 people, two-thirds of whom live on the principal island of Tongatapu. Natural hazards include tropical cyclone, drought, earthquake, *tsunami* and volcanic eruption. There are five islands which are active or dormant volcanoes, and two of these are among those permanently inhabited. The island of Niua Fo'ou is one of these, situated at the extreme north of the island group, almost 400 miles from Tongatapu and the capital Nuku'alofa; 105 miles west from Niuatoputapu, the nearest island, and 215 miles north-west from Vava'u, the nearest and northernmost island sub-group.

Their location at the extreme north of the island group places the islands of Niuatoputapu and Niua Fo'ou roughly midway between Fiji and Western Samoa. From the times of the earliest sea voyagers, these islands became an important 'halfway house' for ships taking on stores and water. The early Dutch and Spanish explorers made use, or attempted to make use, of these islands in this way, as they had been for at least 1500 years previously by ocean voyagers, not only between the countries of Fiji and Samoa but between the two civilizations of Micronesia and Polynesia. This strategic location has probably accounted in the past for the comparatively large populations of each of the two islands; the ships of the Dutch explorer Le Maire were attacked in 1616 by 1000 men from Niuatoputapu, and later by 'warriors in 14 canoes' from Niua Fo'ou.

Niuatoputapu is seven square miles in area; Niua Fo'ou is 19 square miles, including a six square mile lake. Niua Fo'ou is volcanic and rises to 588 feet (see Figure 1). It has no natural harbour and for many years was called 'Tin Can

* Formerly published in *Journal of Administration Overseas* Vol XVIII No 2 Ministry of Overseas Development/HMSO April 1979.

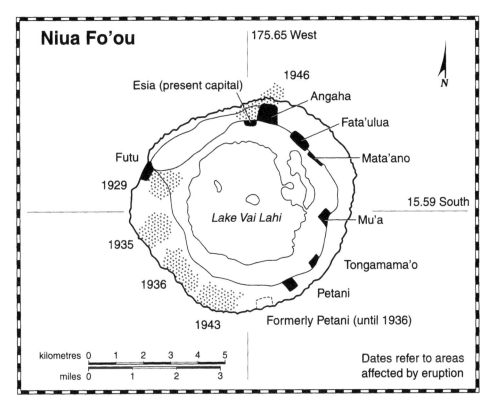

Figure 1

Mail Island', because mail was dropped from passing ships in sealed tins and retrieved by swimmers. In 1946 the population of Niua Fo'ou was 2500.

It would be interesting to know what the effect of earlier eruptions in history has been on this island's large population, but records have been traced back only to 1929, when Niua Fo'ou erupted in July of that year and the village of Futu and many houses and plantations were destroyed. The need for improved communication was recognized by the Government as a result of this eruption, and a wireless station was established soon afterwards. Eruptions came not from the peak but from surrounding slopes, to the sea. They occurred again in 1935 (this time without damage); in 1936, destroying the village of Petani; again in 1943; and then another major eruption occurred in September 1946, as a result of which, although there was no loss of life, damage was extensive. The wireless station was destroyed, as were other government buildings, dwellings and plantations, for the second time during Queen Salote's reign. The decision was made by the Queen and her Government to evacuate the whole population in two stages, first to Tongatapu and then to the island of 'Eua.

In 1977 it was the island of 'Eua that suffered greatest damage from earthquake. In the following year the Government of Tonga requested technical assistance and it was during this assignment that a visit to 'Eua to inspect earthquake damage provided the opportunity to seek out whatever recollection was available of the eruption and evacuation of 1946.

Tongans generally are traditional masters of oratory and written documentation, and Moeake Takai, Town Officer at Angaha on 'Eua is no exception. Aged 20 at the time of the eruption, he had always kept a diary (and still does), and it was with reference to his diaries that he described the events of 30 years earlier, sitting in front of his house in the moonlight.

Darkness falls early in Tonga and there was a full moon on the evening of Monday, 9 September 1946. In Angaha, the principal village on the island, there was a brass band practice at the Catholic Mission, and Moeake Takai and his friend John Malekamu went to football practice. At 7.30 pm they felt the first earthquake, then there were two more and a total of probably ten or 12 minor tremors. The 14 people at band practice stopped playing and, in fear of an eruption, went to their homes. (Three years previously there had been an eruption on the far side of the island after only one earthquake, and on that occasion people had stayed where they were.) Twenty-five minutes later, at five minutes to eight, there was a loud roar from the western side of the village.

Moeake Takai ran home and John Malekamu, who was operator at the government wireless station, reported for duty. Moeake Takai's mother and father had left their home and had joined others in climbing up the mountain away from the village on the coastline. Moeake collected his wife and did the same. People were crying and praying, calling to one another and shouting to find one another. There was a great deal of noise. Eventually they looked back but their village, the principal village on the island, had all gone. Where it had been was engulfed in black smoke.

Climbing up the mountain from the village, as instructed by the police and the district officer, facilitated escape from likely eruptions nearer the coastline. By eleven o'clock at night everyone was on the mountain, including the district officer, the three police officers and the doctors, but with the exception of the wireless operator. As moonlight shone golden on white steam and black smoke, ministers of all churches conducted prayers together with the people, government officers, leaders and chiefs.

Eventually, at about three or four o'clock, the eruption ceased and Moeake Takai went down to look for his friend, the wireless operator. John Malekamu had sent out SOS signals for help and had left the wireless station only when lava was 30 feet away just before its destruction. Exhausted by the intense heat, he had run and finally collapsed under a tree, where he was eventually found by his friend. Having realized that whether he lived or not was now 'up to God', he survived and was taken to join the rest of the island community up the mountain. He was able to report to the police the extent of destruction in the village. At sunrise all joined in giving thanks to God for their deliverance – without casualty.

Among the 2500 people on the mountain sides were four Sisters, one each from Belgium, France and Holland, and one from the USA who had with her her national flag, which was raised on the highest tree to attract attention. The eruption had commenced on the Monday evening and all the island community was gathered on the mountain sides by Tuesday, the following day. They remained there throughout Tuesday, Wednesday and Thursday; and on Friday, 13 September, at ten o'clock in the morning, a plane was heard. A US Navy plane flying from Pango Pango in Samoa to Na'adi in Fiji saw smoke still rising from the island and the waving of clothes on the mountain sides, and reported the situation at Na'adi. A message was sent to the Tonga Government at Nuku'alofa where none of the SOS messages from Niua Fo'ou itself seemed to have been received. The people remained where they were throughout Friday, and until Tuesday, 17 September, when a government ship was at last sighted, one week and a day after the eruption. A cargo ship also arrived on the same day. At 4.30 in the afternoon a government officer arrived on the mountain to ask the people to move down again.

A meeting was arranged between government officials who had arrived from Nuku'alofa, the Chief of Niua Fo'ou who had been in Nuku'alofa, and leaders of the island communities. On the following day a public meeting was arranged, where a government official expressed the need to evacuate the entire population to Tongatapu. The island communities are divided between four chiefs, and when the time came to vote on the issue of whether to agree to government evacuation or not, three chiefs and their communities of 1800 people voted in favour of evacuation and one chief and his 700 people voted against evacuation. Tonga has had a democratic form of government since 1875, however, and the majority vote in favour of evacuation meant evacuation for the whole island population of 2500 people.

Saturday, 21 December was fixed for the commencement of evacuation to Tongatapu. The police magistrate from Nuku'alofa remained at Niua Fo'ou as chairman of the Evacuation Committee and other officials returned to the capital. Although there was, therefore, the considerable time of three months to prepare for evacuation, there was, on the other hand, a lot of work to be done. All surviving dwellings were to be taken down to permit reuse of building materials, and all material had to be stacked, ready for loading and embarkation.

When the day came, loading began at six o'clock in the morning and continued until five o'clock in the evening. The 2500 people and building materials were transported to Nuku'alofa in one day. No animals were permitted to be taken, no food, and no sewing machines – only suitcases for clothing and personal belongings. Each person had been allocated a number, and each number was checked as embarkation progressed. It was an efficiently ordered occasion, but many people were crying and very upset. There had been, in fact, two eruptions, one adjacent to the principal village on land and the other just offshore. The landing place between these two eruptions had been totally destroyed. Three boats came from Nuku'alofa for the evacuation. One was loaned by the US Government and another, the Matua, came from the Union Steamship Company of New Zealand. These two ships had to anchor three-quarters of a mile away from the shore and the third boat, a small wooden craft, the Hitofua, was used to ferry all passengers and goods from shore to ship.

Twenty-one men and one woman were left on Niua Fo'ou to tend crops and what property and belongings remained, but they were there only from December to the following April. Life for them must have been particularly difficult, and they were eventually taken off by a Seventh Day Adventist schooner to Samoa. From there they were taken to Nuku'alofa, still in the charge of the minister on the boat to Vava'u. The woman was taken ill and died on board.

Tongatapu was to be only a transit stop. In September 1948 the move began to the island of 'Eua. The distance from Nuku'alofa is only 25 miles, but only the contingent of 1800 people who had voted, with their three chiefs, in favour of the evacuation agreed to go on to 'Eua. The other 700, with their chief, stayed in Tongatapu until 1959, when their requests to return to Niua Fo'ou were recognized, and the government agreed to assist their return.

The island of 'Eua, with an area of 34 square miles, is one of the largest in the Tongan island group, but even now has only 4000 population, 4 per cent of the national total. The island is mountainous, rising to just over a thousand feet, naturally forested and with fresh water springs. 'Eua is larger, more fertile, and nearer to the capital than Niua Fo'ou but the first ten years were a very difficult time for the newly-arrived population. Agricultural land had to be cleared and crops and planting material had to be established. Forest timber, in government ownership, was not freely available and there were few coconut palms available for building. In 1957, a new sawmill was established with New Zealand development aid, and a forestry scheme was established, under which one quarter of all timber felled on farmed land was made available to the allotted occupier. This marked a turning point in the island's relative prosperity and standard of living.

In the intervening ten years it had probably been severe economic hardship in unfamiliar surroundings that caused 74 men to sail to Niua Fo'ou in 1950 to work on copra production. They stayed on the island for one year, and then returned to 'Eua, being replaced by another working party in 1951, out of which 24 men stayed. In 1952, about 20 men stayed, and again in 1953. Moeake Takai joined one of these working parties and briefly reflected on the moving experience of a return to his native island. Animals had been untethered and freed by the group of 20 who stayed in 1947, and had since gone wild. The horses, however, responded to the long-forgotten sound of water buckets, and allowed themselves to be mounted to join in the work of coconut collection.

There was no permanent settlement on Niua Fo'ou from 1947 until 1959 and all land reverted to government ownership. It was probably the work of the copra working parties that encouraged the government to assist the permanent return of the contingent from Tongatapu. Copra is the principal export commodity of Tonga as well as the principal cash crop of the majority of islanders. In 1977, in spite of government assistance for the return of 'the seven hundred' from Tongatapu, land on Niua Fo'ou had not been allocated by central government under the Tongan tax-allotment system for life tenancy by farmers.

The indications are that the evacuation of 1946 was *ad hoc* and unplanned, and that the decision to carry it out was precipitate. The apparent lack of forethought about the ultimate relocation of evacuees must have made their plight far more onerous than it need have been, and their ultimate resettlement a much longer process than necessary.

Tonga gives serious attention to the environmental hazards by which her people are afflicted. The 1977 measures for disaster mitigation recommended co-ordinated and integrated preparedness measures, attention being drawn to the significant aspect of contingency planning for the evacuation of populated volcanic islands. Preparedness planning indicated the need for policy decisions concerning temporary or permanent relocation of communities as part of a context of development planning, in recognition of the need to arrest spontaneous movement of people from outer islands to urban centres. Infrastructural and administrative development of Niua Fo'ou was planned to provide an airstrip by 1980 and a six-bed in-patient ward for the dispensary. The absence of secondary schooling, commented upon in the Third Development Plan, is a significant cause of migration to larger islands. Additional services such as these will reinforce the island's resources in the event of major disaster, but local and national disaster preparedness, serving to co-ordinate and generate infrastructural development, may do more to remove resistance to living on an extremely remote and hazardous active volcanic island.

References

Rogers, Garth, (Ed) (1986): *The Fire has Jumped: Eyewitness accounts of the eruption and evacuation of Niuafo'ou, Tonga* Institute of Pacific Studies, University of the South Pacific, Suva, Fiji.

CASE-STUDY II

Some Perspectives on Natural Disaster Vulnerability in Tonga*

THIS CASE-STUDY is an edited version of the original* which was based on the Report of a Technical Assistance Assignment undertaken in 1978 on behalf of the Overseas Development Administration (now the Department for International Development), London, for the Government of the Kingdom of Tonga. The author wishes gratefully to acknowledge permission granted by the Government of Tonga for the original publication.

Located approximately 725 kilometres south-east of the main islands of Fiji, and 1930 kilometres north-east of Auckland, Tonga consists of a total of 172 islands, 36 of which are inhabited. Administered from the capital, Nuku'alofa, on the principal island of the Tongatapu group, the archipelago includes three other sub-groups; the three islands of the Niua group in the far north, the Vava'u group and the Ha'apai group. The overall distance between inhabited islands from north to south is 690 kilometres. The total land area of Tonga is 730 square kilometres; total populated land area is 647 square kilometres; and the preliminary estimated total population is now near 100 000 (Figure 1).

It remains the case that most work to do with natural disasters deals with single disaster occurrences; some may refer to disasters collectively in a number of different places, and other more mono-disciplinary studies may examine hazards of a single type throughout a period of time in a particular place or region (Lewis, 1977). Rarely have studies departed from considering each disaster or disaster type, to offer analytical perspective related to recurrent disasters in a particular location or country.

An exception was the multi-disciplinary study of hurricane hazard included in a supplementary report to the study of population resources and development in the eastern islands of Fiji in the Man and Biosphere Programme (UNESCO, 1977): and Glass *et al.* (1977) made incidental but significant reference to the importance of historical changes in dwelling construction between two earthquakes affecting a Guatemalan village in 1918 and 1976. An early study in pre-disaster planning examined the history of hurricane impact in the Bahama Islands (Lewis, 1976) and a regional study in the South Pacific examined all major environmental hazards for a nine-country region (Lewis, 1976).

Study of the impact over time of all kinds of disasters in one location commenced with the study of 'all hazards at a place' – the place being London, Ontario (Hewitt and Burton, 1971). Such 'in place' studies are realistically illustrative of the continuous indigenous experience of hazards.

An assignment in Tonga (see above) provided further opportunity to examine

* Formerly published in *Pacific Viewpoint* Vol 22 No 2 pp 145–162 Victoria University of Wellington. 1981.

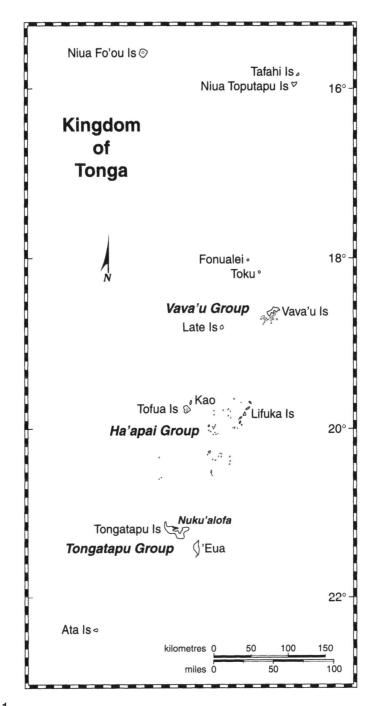

Figure 1

61

the history, where records permitted, of all hazards and disasters in the Tonga Islands (Lewis, 1978). Additionally, a part of the history of these events, which have played their part in moulding the physical and cultural nature of the islands and their populations, has been the more recent history of post-disaster assistance. Some analysis of needs proved helpful in the identification of preventive and relief measures to be taken for future disasters. The assignment was undertaken exactly a year after the earthquake of 1977 and within a few months of the end of the serious drought and two severe hurricanes, Anne and Ernie, in the same year and in January 1978, all of which provided material to place into longer perspective and comparative analysis.

Disaster history

The appearance, or reappearance of Fonua Fo'ou (new land) renamed 'Jack-in-a-box Island', recorded in 1781 by the explorer Maurelle (Rutherford, 1977) is one of the few historical references to hazards or disasters in Tongan history from before the mid-nineteenth century. Fonua Fo'ou had disappeared again in 1978, to await another seismic upheaval before reappearing.

Over the Tonga archipelago, in the 100 years between 1875 and 1975 there were recorded 28 hurricanes, 22 earthquakes of moderate magnitude or greater, five periods of drought, four volcanic eruptions and three known *tsunamis*; a total of 62 recorded events or a national average of one for every one-and-a-half years (Lewis, 1976).

Some of the world's largest earthquake magnitudes have been recorded in Tonga, such as the 1919 earthquake (Richter 8.4; Richmond, 1976) and that of 1917 (Richter 8.7) which raised the floor of the lagoon on the island of Niua Toputapu and caused it to dry (Angenheister, 1921).

In addition to Fonua Fo'ou, there are five active volcanic islands in the Tonga group, two of which are inhabited and one of these, Niua Fo'ou, remains only partly inhabited after the evacuation of 2500 people after the eruption of 1946 (Lewis, 1979b and Case-study I).

Cyclonic storms have nevertheless been the most frequently damaging hazard, those achieving sustained wind speeds of over 120 kilometres per hour being classified as 'hurricane'. Several islands may be seriously affected by one hurricane, as may several countries. Hurricane Bebe of October 1972 caused catastrophic damage on Tuvalu (then the Ellice Islands) and Fiji before striking Tonga and then causing lesser damage on Niue. The scale of impact on these very small countries is colossal and overwhelming. Ninety-five per cent of all houses in Funafuti (Tuvalu) were totally destroyed in 1972 and a massive section of the reef itself was shifted by the sea (Baines and McLean, 1976). Twenty-two per cent of the national population of Fiji was rendered homeless (Lewis, 1979a and see also Lewis, 1983b for damage caused throughout the archipelago by Hurricane Isaac in 1982).

History provides only a short record by comparison with Tongan prehistory. Polynesians were sailing the ocean 1500 years before the arrival of the first

European explorers (Rutherford, 1977). It can be assumed that earthquake, volcanic eruption, hurricane and *tsunami* were as evident in prehistory as they are in history, but history is recorded principally by Europeans. Otherwise descriptively rich Tongan folklore and legend seems unconcerned with reference to environmental hazards, with the possible exception of volcanic eruption*. Were environmental hazards simply a fact of life (and death) to be accepted and absorbed in much the same way as night and day? Dr Garth Rogers (University of Auckland) has confirmed that in a language as poetic and as rich as Tongan, it would be unimaginable that there should be no word for a phenomenon, natural disaster, which visits so often. The concept is expressed as *fakatu'utamaki* but is one which history and prehistory have moulded into the normality of Tongan life. With this perspective, when did the concept of 'disaster relief' first appear, and what were the conditions that made its apparent need so different from traditional responses?

First references to relief come from the Colonial Report of 1909 (Westgate, 1975) – another product of the writing of Europeans. After Niua Fo'ou was hit by hurricane in 1909 'the Government of Tonga sent in relief but it was not required to any great extent'. Does this brief reference suggest compliance with an alien concept by an essentially alien administration for which indigenous populations were unprepared? Certainly, in the brief notes taken of the colonial reports by Westgate there is no further reference to 'relief' until the need for emergency shelter for 8000 people is mentioned after the hurricane of 1961 in the Vava'u and Ha'apai island subgroups.

In the interim, there had been 13 hurricanes or severe storms, six periods of drought, two volcanic eruptions and one 'strong' earthquake, deemed worthy of mention in the colonial reports, during the 40 years during which Reports were uninterrupted by World Wars. Twenty-eight other occurrences were recorded in other sources (Lewis, 1978), making a total of 50. By contrast, in all but four of the total of 23 events in the colonial reports, the impact on production, particularly of coconuts or copra, receives significant mention. This was the priority concern of an essentially colonial government.

With the exception of the 1946 evacuation of Niua Fo'ou following the eruption, and the 'considerable property damage' of the 1961 hurricane, 'disaster' damage to the indigenous population was ignored in the reports. Was this because it was effectively absorbed by the indigenous population, or because the administration was insufficiently interested, or a combination of these?

It was not possible to find any official record there may have been in Tonga of the 1961 hurricane, but by common consent among representatives of Tongan central and provincial government, 1961 represents the point at which significant disaster relief first appeared in response to damage caused. In that year, 50 per cent of housing in the Vava'u and Ha'apai sub-groups was destroyed or badly damaged, banana crops were wiped out and coconuts stripped and uprooted,

---

* This statement is based on cursory reading and was corroborated by The Hon Ve'ehala, Acting Minister of Works and Education, Governor of Ha'apai, Cabinet Minister, and Chairman of the Tonga Traditions Committee.

and it took two and a half years before the next shipment of copra was despatched (Kerr, 1976). What emergency shelter was provided, if any, is not known, but whatever was done was, again by common consent, far less than what has been done since and recently in similarly serious hurricanes in other parts of Tonga (Lewis, 1978).

After Tonga became totally self-governing in 1970, hurricane Juliette of April 1973 was the first severe hurricane, as a result of which, the Hurricane Relief Committee was established by the Prime Minister. Seventeen villages in seven islands of the Ha'apai sub-group were seriously affected and 1250 families required rehousing; that is, 8000 people, three-quarters of the population of the whole Ha'apai group and 10 per cent of Tonga's national population. Food, water, and planting materials were provided by the Committee from central government, and later a construction programme for the 1250 families, with financial participation from each household, was commenced.

Official reports are not available of the effects of hurricane Juliette. An appeal for assistance was addressed to the New Zealand High Commissioner by the Minister of Works, the Honourable Langi Kavaliku in November (Kavaliku, 1974). There is a significant shift of emphasis by comparison with colonial reports. In a four-page letter, although 'damage to crops' received a passing mention, coconuts and copra production are not referred to at all. The emphasis throughout the Minister's letter, in addition to that of immediate relief, is on necessary improvements to normal infrastructural resources so that conditions following disasters may he better attended to. The letter struggled to separate the problems caused by hurricane from the everyday inadequacies that the hurricane had exacerbated. It also clearly identified the main aspects for concern as being food, shelter (rebuilding), water supply and transport communication.

By the time of hurricane Anne in December 1977, hurricane disaster reports had become well ordered and highly detailed documents (Central Hurricane Relief Committee, 1978). Using the MV *Kao*, a front-loading ocean-going barge given by the New Zealand government as a result of the appeal after hurricane 'Juliette', a team of five representatives of the Hurricane Relief Committee headed by the Minister of Health, visited all 16 islands of the Ha'apai sub-group, a round trip of eight days. Detailed assessments of damage were made in three clearly defined sections of agricultural food supplies, water sources and supply, and government buildings and private housing. In each section, the need for improvement of normal conditions, for the purposes of mitigating future disaster damage, is clearly and overtly stated.

The section on water sources and supply estimated damage as T$1534 but proposed the sum of T$40 836 as being necessary 'for improvements'. The section on government building and private housing observed that: 'Most of the church buildings and private houses that were damaged or destroyed were very old and should have been pulled down and (or?) repaired long before the hurricane'. The report concludes with recommendations to cabinet which include a national and international appeal for hurricane relief aid, financial concessions to ease the cash burden on households, an allocation of priority for the repair of government and community cisterns, and that three wharfs damaged by the hurricane should be repaired.

In the 12 or 15 years after the hurricane of 1961, there was a marked change of emphasis from reductions in copra production, to the basic needs of indigenous communities. What is also significant is the change from a situation where not only government ignored the concept of relief, but one where, when it was first offered in 1909, it was not required.

In recognizing the clear relationship between normal conditions and disaster damage, were the ministers of the Hurricane Relief Committee quick to seize upon relief aid, which was readily available for the asking after disaster, to augment their basic needs development programmes? (After the earthquake of 1977, relief aid came into Tonga from 15 separate governmental donors and seven different non-governmental sources.) If they were, then clearly they were not covert in doing so. Their reports are logical, clearly stated, direct and available. As representatives of a tiny nation in the middle of a massive ocean, they are at the mercy not only of the elements but of the international relief- and development-aid machine. By not seizing an opportunity to augment their development budget they would not be serving their country and people as well as they might. Is this the purpose of relief aid? Should not relief needs be identified more closely so that post-disaster assistance can be more sure of its purpose and of its effectiveness?

Vulnerability

Attempts to discern consistent regional variations of risk of the occurrence of hurricane and earthquake within Tongan islands were inconclusive (Lewis, 1978). The result of an analysis of a number and variety of previous studies and records confirmed that, although there is periodically an apparent emphasis of risk for one island sub-group or another, overall, over longer time periods, no sub-regional allocation of specific risk could safely be made. Recommendations were made for improved methods of climatic and seismic monitoring to be established so that regional locational vulnerability could be reassessed.

Population distribution within the island group is therefore the first indicator of vulnerability to disaster. With an even spread of risk of occurrence assumed for the time being, where concentrations of population are greatest, vulnerability to loss is also greatest. Movement of population, as well as any natural increase, may therefore lead to the exacerbation of that vulnerability.

In 1921, a Professor Angenheister of the observatory at Apia in Western Samoa wrote:

The present paper deals with those [earthquakes] whose epicentres are near that part of the ocean known as the Tonga Deep. Fortunately, although these earthquakes are numerous, there are few inhabitants in that part of the Pacific, and consequently, the earthquakes have little destructive effect, more especially as most of them are under the sea (Angenheister, 1921).

The Tongan population in 1921 was 24 935, one-quarter of what it is now (Kingdom of Tonga, 1975). The population of the largest island, Tongatapu, on which the capital of Nuku'alofa is situated, has increased sevenfold since 1921. Natural increase and migration (inclusive of the evacuation of 2500 people from Niua Fo'ou in 1946, 1800 of whom went on to 'Eua) thus have a dual effect on vulnerability assessed in this way. Locational vulnerability may be exacerbated by 'vocational vulnerability' – activities which affect the three basic elements of life support, being water, food and shelter.

Rainfall in 1977 was considerably lower than the annual average for the preceding eight years (Kingdom of Tonga, 1975) and drought prevailed for three months, from October to December. However, more than half of the year's rainfall fell during the first three months. Most Tongan atoll islands do not contain fresh-water lenses within their porous coral-rock formation. Other islands are raised to higher elevations, but have hitherto resisted the drilling of wells due to their height and hardness of rock structure. Throughout all islands, drinking water supplies are from roof catchment collection and storage tanks or cisterns on and in the ground (see cover photograph). The last major government building programme for the construction of cisterns for water storage throughout the Kingdom was undertaken in 1908–9 (Rutherford, 1977) when the total population was 22 000.

There is piped water supply in the capital and in some principal villages, but elsewhere increase in population has caused more intensive use of storage systems. Rainwater was once collected off several roofs and conveyed by pipe or gutter to the cisterns. Now in many cases, roofs have disappeared with the buildings themselves, and cisterns are often fed only by the water drained off their own roofs! More recent buildings with large roofs, for example schools and churches, whilst collecting water for their own use in private tanks, have ignored communal needs and communal cisterns.

It is not only drought that is exacerbated by inadequate water catchment and storage. Environmental health hazards are more likely to increase in conditions of water shortage and the consequent decrease in standards of personal hygiene. Deprivation will be more widespread where water storage content and capacity is low, following hurricanes and storms which may have damaged roofs and guttering systems. In July 1978, a concrete cistern capable of holding 68 000 litres held barely 40 millimetres of water to serve an isolated island community of 350 people.

In such conditions the liquid from young coconuts provides the only source of thirst prevention, and some consumption of coconuts for drinking and eating purposes is normal throughout the Pacific. In times of water shortage and drought, however, the consumption of coconuts might increase up to four times, with consequent and corresponding decrease in coconuts available for subsequent copra production. Estimated coconut consumption during the three month drought of 1977 was equivalent to 11 per cent of the copra production of the Ha'apai island sub-group (Lewis, 1978). Copra is the principal source of cash income for Tonga and Tongans, who cannot afford to have their source of income eroded in this way. Capital expenditure on improved systems of water catchment and storage would be quickly compensated for by maintaining copra production.

The need to produce copra for cash is increased by the endemic consumption of coconuts for food and drinking, and the need for cash cropping generally is similarly increased. Over-emphasis on cash cropping, however, to the exclusion of food crops, increases vulnerability to disasters. Although food crops are themselves vulnerable to hurricanes, staple root crops of taro, cassava, yams and sweet potato remain edible for some time after their foliage has been destroyed. Although damaged, therefore, as long as these crops (and others) are present in adequate quantities, there will be food to eat for up to two or three months afterwards. The reduction of food crops in favour of cash crops – of vanilla for instance – reduces capacity for self-reliance and survival. The availability of cash in lieu serves little purpose when food supplies for market are unavailable as the result of scarcity brought about by cash crops, and additionally by the devastating effects of hurricanes.

Priority for cash cropping in the most favourable and fertile agricultural sites causes the relocation of food crops to marginal ground, which may increase the vulnerability of food crops to damage. The process was described in the 'Study for the Man and the Biosphere Programme' in the eastern islands of Fiji:

> Since the best soils are on the coastal flats and in the lower valley bottoms, increasingly the most favoured areas have been used for coconut plantations. . . The gardens have had to extend up-slope and up-valley into areas formerly regarded as too steep, difficult, poor or remote to be cultivated. . . These slopes are, however, very exposed to wind and storm damage. In 1975 a larger proportion of the crops was destroyed by the hurricane or subsequently rotted in the ground than was the case in 1948. . . (before the process described had become so well established). . . (McLean et al., 1977).

Traditional building construction techniques in Tonga, as in most other less-developed countries, have gradually given way to the adoption of Western materials and methods. Corrugated iron sheeting and sawn timber frame construction have been practised in Tonga since colonial times, and sheet roofing has in many instances replaced thatch in traditional Tongan *fale* construction. (The more widespread use of sheet material for roofing makes water catchment more effective and systems for water storage more beneficial.) Subsequently, the use of concrete blocks has increased, together with the use of reinforced concrete for domestic construction, as well as for government, commercial and religious buildings.

The whole of the estimated cost of damage in the 1977 earthquake, slightly more than one million dollars (NZ), was for building damage (including wharfs) and attention was drawn to the fact that buildings most severely damaged were those built of modern materials, reinforced concrete and/or blockwork (New Zealand Ministry of Works; Campbell, M.D., 1977; and New Zealand Department of Scientific and Industrial Research; Patterson, 1977). Traditional *fale* buildings were not damaged, nor were the 'sawn timber' buildings, except in some cases by failure of raised masonry foundations or by broken windows. It is obviously apparent that earthquake damage increases with the unregulated increase in the use of non-traditional building forms and methods. In a future of increased development activity at all levels, this increase in future earthquakes will be compounded.

The 1977 earthquake exposed abysmally low standards of construction where modern materials had been used. Reports noted widespread damage in an earthquake of 'low to moderate' magnitude (7.7 Richter) having a return period of 13 years. The New Zealand reports emphasized the crucial importance for Tonga of building regulations to ensure improved standards of construction and supervision.

Legislation for improved standards of building construction would not be adopted lightly, however. Construction standards are so low that any general legislation, even if practicable, would impose severe cost burdens on the industry. In the meantime, all available means of instruction will need to increase awareness of earthquake hazards and appropriate construction methods – through loan authorities, government labour training programmes, short courses for the commercial sector, leaflets and posters, and schools. Meanwhile, it is understood that those who turned to blockwork and concrete construction following damage to wooden houses in hurricane, having now suffered most heavily in the earthquake, are rebuilding once more in timber. Opportunity for a governmental lead, in the cycle of destruction and deprivation, for improved methods of timber construction resistant to 'all hazards at a place' is now open.

Relief and development

The Hurricane Relief Committee reported its recognition of the need to improve normal conditions, if resources and capacity for indigenous disaster mitigation were to be increased. The report of the ODA Technical Assistance Assignment, among other things, identified some of the processes that have led, or may lead to, increased disaster vulnerability. It has been suggested that there are serious shortcomings within international aid, both development and relief aid, which seeks either to improve 'normal' conditions, or to ameliorate disaster, or both.

The most obvious products of development aid in Tonga are in the construction of wharf facilities for tourists and goods, hospital and school construction, commercial fishery development, and developments in agriculture for animal husbandry and forestry (Kingdom of Tonga, 1976). Commercial enterprise has provided a satellite communication earth-station, an extended airport, and has attempted to establish an oil extraction industry. In addition to training programmes and advisory and technical assistance programmes, the emphasis of development aid has been on capital expenditure on projects of a physically identifiable kind. Is it in the nature of current policies of international aid that results must be visibly recognizable and therefore more readily identifiable with their donor agency?

This kind of aid has led to a burden of maintenance and recurrent costs which falls on the national government and has to be paid for out of the national budget. Building and infrastructural maintenance, and the provision of small buildings, additions and alterations to capital-intensive projects, place a burden on national resources which leads to a run-down of maintenance generally.

This may explain the reasonable temptation to direct funds from relief aid to small-scale building improvements as well as repairs. There would seem to be no fault in this as far as Tongan administration is concerned. The lower the standards of maintenance in buildings or infrastructure, the greater the likelihood of damage and consequent loss in hurricanes and earthquakes. Nevertheless, the cart is before the horse. With more development funding for maintenance and small-scale projects, the greater will be national capacity for disaster reduction.

Programmes of development expenditure require a widespread multiplicity of small projects; one possibility (identified above) is in water catchment and storage systems. 'Improvement' is clearly synonymous with 'development' in this context.

Another recommended programme is the development of commercial fishing, probably undertaken through village co-operatives and the Development Bank. The improvement of local fishing industries provides not only an increase in alternative food supplies, but by its infrastructure provides the means for locally initiated inter-island exchange – the traditional life-blood of survival in disaster. Spontaneous assistance from unaffected islands to those that have suffered, is a Polynesian system of extended mutual aid which has become eroded and largely destroyed by over-emphasis on centralized government and communication in the past.

Whether development programmes are devised under budgets for disaster reduction, or whether they remain within basic needs development, is immaterial. What is important is that they are implemented, and that their role in disaster reduction is recognized by both the donor and the recipient authority.

Against a background of long-term accretion of vulnerability to natural disasters, the need for post-disaster assistance is not disputed. What has to he examined is the form that assistance takes and its integration into short- and longer-term uses and effectiveness.

It could be said that natural disaster is the monitor of development. What happens in disaster is the manifest failure of development not ideally socio-environmentally tuned. At a lower scale, disaster relief may be the indicator of localized vulnerability. Effective analysis of relief needs would identify vulnerable locations or social groups. Where needs are not effectively identified, relief may he useless. The identification of localized or sectoralized vulnerability requiring relief in onc disaster should be the signal for redirection of development aid for 'improvement' in time for subsequent disaster of the same or different kind. Development and relief aid must be linked, not only within donor organizations, but within national development administrative systems of disaster-prone countries. With recognition of the processes which may lead to vulnerability at local levels, development at local levels can be directed accordingly.

Examples of 'useless relief' are a part of folklore and are held in perpetuity. In Tonga after the drought, and hurricanes Anne and Ernie in 1977 and 1978, local expatriate officials of an international non-governmental organization appealed for vitamin tablets to their parent body. A representative was sent to determine the veracity of the appeal but, by the time the representative's report was received, one million vitamin tablets had been despatched. Despite the reported conclusion that victims of the hurricane were nutritionally fit and healthy, and despite Tongan Ministry of Health agreement to those findings, distribution of the tablets ensued. With sufficient tablet supply for three months, the burden on local administrative resources was colossal. Tongan cultural tradition not to give offence by refusing a gift prevailed. Post-disaster assistance placed a giant burden on administrative infrastructure at a time of emergency when those indigenous resources should have been more effectively used on matters of higher priority of need.

Some relief aid may have negative effects where it is not offered in specific response to locally identified need. The availability of such relief supplies may create future artificial needs, or dependency on them. Dependency may be psychological or practical and is often difficult to identify. Is the distribution of canned fish from Japan – another gift – a result of a shortage of locally caught fish, or is the local inadequacy of fishing partly a result of the easy availability of canned fish? Does the monthly distribution of flour, sugar and dried milk over nine months – gifts from Australia – have a bearing on feckless behaviour in local garden management? There are numerous Tongan administrators who would say that it does.

Dependency comes in many forms, not all of which can be blamed on disaster relief, and not all of them present in Tonga. Dependency on cash, cash cropping, the need to work in paid employment, consequent migration, depletion and mismanagement, is a more serious dependency cycle – which may exacerbate disaster vulnerability to a far greater degree. But the relationship of disaster relief to these negative aspects of 'development', as well as to the positive aspects, must be understood if integration of relief aid with development aid is to succeed.

Whatever the causes and origins of dependency, vulnerability can only increase as a result. The greater the degree of dependency, the less is the capacity for self-reliance. In disaster, dependency is exposed through the disappearance of the dependency prop, in cases where former systems for self-reliance have deteriorated due to a hitherto dependent condition.

Examination of the effectiveness of relief aid might reveal some of the negative by-products, might create opportunity for longer-term presence by executive representatives of governmental and non-governmental donor representatives, and might reveal opportunities for continuing involvement by donors of relief aid in programmes of basic needs development that would reduce disasters. Where this already occurs, however, due to the presence of some representatives, it does so apart from, and unrelated to, the major development programmes. There is not only a lack of integration of relief and development aid in this sense, but also between small-scale and large-scale development projects and programmes. Integration is a mutual and two-way process. Once again, the pragmatic redirection of relief-aid funds into small-scale improvement measures can easily be understood in this context. There needs to be a breaking of the cycle of relief–dependency–vulnerability–relief, and exposure of inherent structural weaknesses that create the need for relief palliatives.

In Tonga it is relevant to distinguish between financial relief aid, which has found an appropriate if not totally logical use, and material relief supplies. It is then relevant to pose the question of whether relief supplies, of the kind there have been in the recent past, are necessary at all. Some cases of destitution and severe need will be found, but it will be necessary to determine whether those cases were caused directly by the disaster in question, or whether they are a product of 'normal' conditions in a poor country. That there is a clear relationship is understood, but it is disaster relief which is here the subject of enquiry.

In common with a number of other countries, especially those of the Pacific, Tonga is well-known for its ceremonial feasting. Rutherford (1977: 78–79) makes a realistic comparison of the social and economic impact of food shortage resulting from hurricane with the social and economic impact of traditional Tongan feasting:

> The land produced for itself a wide variety of plants and trees. Some agricultural products were cultivated: yams, bananas, and plaintains for food; and the paper mulberry and the pandanus for cloth and mat making . . . *Kava* was cultivated for the use of the chiefs and for ceremonial purposes. Pigs and fowls, the only domesticated animals, were left to scavenge for food and to breed at will. Although the Tongans were excellent farmers, aided by a rich soil and favourable climate, famines were frequent. They were caused on occasions by unusually dry seasons, or by devastating hurricanes. More often than not, however, shortages of food resulted from excessive consumption of food at *inasi* ceremonies, weddings, funerals and voyages to the outer islands. At the marriage of the Tu'i Tonga, for example, food was stacked in heaps, sometimes fifteen or eighteen metres high (corroborated by reference). Some attempts were made to lessen the frequency and severity of famine by declaring certain foods *tapu* following large-scale feasting. The *tapu* was also employed before some anticipated ceremony or festival to ensure that ample food was available at the appointed time. At these times of *tapu* the common people suffered and were driven to seek edible plant roots in order to survive. There appears to have been some concern to over-produce in order to meet the demands of ceremonial and obligatory presentations of food, but only limited attempts were made to store food for lean times, the exception being the preservation of bread-fruit in pits and storage of yams in specially constructed shelters.

The impact of feasting on the consumption of food is still recognized, and as recently as the drought in 1977 a government circular was issued to District and Town Officers on how not to waste food. Feasting was not to be *kai pula* but *kai peleti* (on plates) and by 'plastic bags'.

Whether 'wastage' is avoided or not, if Tonga can absorb the extremes of feasting to the considerable extent it does through social coping mechanisms, then it is conceivable that it may have similarly absorbed the impact of environmental extremes in the past, of similar impact and lesser frequency, and could do so again in the future.

An item in the *Tonga Chronicle* of June 1977, under the heading of 'New Zealand Aid for Ha'apai' reported a cheque for T$24 000 given by New Zealand to the hurricane relief programme. Indented within the same column space was a short account of the royal feast given to 800 guests in honour of the King's 60th birthday:

A spokesman from the feasting committee said 1462 suckling pigs and 1223 chickens were roasted for the royal occasion. There were also enormous amounts of seafoods and other delicacies.

At market prices of T$15.00 per suckling pig and T$3.00 per chicken, the cost of those two items alone would have exceeded the relief donation from New Zealand by T$1600.

In the Government of Tonga in 1975, the Central Planning Unit was created within the ministry of finance as part of the preparation of the Third Development Plan 1975–1980: later it became the Central Planning Department. The report of the Technical Assistance Assignment recommended that all matters pertaining to development (i.e. improvement) arising from the occurrence of of natural disaster should be handled by the Development Planning Department. The report also suggested that the Department may become more closely and effectively able to plan to take environmental hazards of all kinds into account. As a further recommendation for institutional matters, it was proposed that the Hurricane Relief Committee should be renamed the Disaster Relief Committee, should continue to have responsibility for relief after all types of disaster, and furthermore that it should take on additional responsibilities for preparedness matters. The Prime Minister of Tonga is chairman of the Relief Committee.

In any country, but particularly in countries comprised of island groups, disaster may be localized in its impact. While none of the 36 inhabited islands of Tonga can be said to be immune to environmental hazards, in any one of them occurrence will be random and without regularity. Nationally, however, the country as a whole has to cope with 'all hazards at a place' – the place being Tonga. Nationally the risk of occurrence can be more certainly assessed, and nationally there is greater opportunity to take account of a more frequent occurrence of extremes. This must not be assumed to imply a sole responsibility by central government for disaster management, but rather a capacity for the initiation of locally managed coping mechanisms planned within national frameworks for the reduction of vulnerability and the administration of integrated relief, where and when necessary. By the regeneration of local coping mechanisms under governmental initiation and supervision, where these have plainly been eroded in the past, local interdependence and self-reliance will probably arrest those processes that have hitherto accelerated dependency on central government – and which have brought about periodic national dependency on international relief aid.

References

Campbell, M D (1977): *Summary Report on 1977 Assignment in Tonga* Director of External Aid, Ministry of Foreign Affairs. Mimeo. Wellington, New Zealand.

Central Hurricane Relief Committee (1978): *Report of a Visit to the Ha'apai Group 4–11 January 1978 Regarding Hurricane Anne*, 14 January. Nuku'alofa.

Glass, R I; Urruthia, J J; Sibony, S; Smith, H; Garcia, R; Pazzo, L (1977): Earthquake injuries related to housing in a Guatemalan Village *Science* 197 pp 638–643.

Hewitt, K, Burton, I (1971): *The Hazardousness of a Place: An ecology of damaging events* University of Toronto.

Kerr, I S (1976): *Tropical Storms and Hurricanes in the South West Pacific: November 1939 to April 1969* NZ Meteorological Service Misc Pub 148. Wellington.

Kingdom of Tonga (1975): *Statistical Abstract 1975: Constitutional centenary number* Statistics Department.

Kingdom of Tonga (1976): *Third Development Plan 1975–1980: Policy and objectives, programme and strategies for social and economic progress* Central Planning Office. Nuku'alofa.

Lewis, James (1975b): *A Study in Predisaster Planning* League of Red Cross Societies/ Disaster Research Unit Occasional Paper No 10. June.

Lewis, James (1976): *A Report to Establish Guidelines for the Management of a Regional Fund to Provide Insurance for Natural Disaster* Commonwealth Secretariat/South Pacific Bureau for Economic Co-operation. London/Suva.

McLean, R F; Bayliss-Smith, T.P; Brookfield, M; Campbell, J R (1977): *The Hurricane Hazard: Natural disaster and small populations* Population and environment project in the eastern islands of Fiji: Islands Reports No 1. Australian National University, Canberra.

Patterson, B R (1977): *Geological Aspects of the Tonga Earthquake: 23 June 1977* Engineering Geology Section EG 297, New Zealand Geological Survey, DSIR, Lower Hutt. Mimeo. December.

Richmond, T N (1976): *Earthquakes and Tsunamis* Paper presented at the South Pacific Disaster Preparedness Seminar, Suva. Mimeo.

Rutherford, I N (ed) (1977): *Friendly Islands: A history of Tonga* Oxford University Press. Melbourne.

The Tonga Chronicle (1977): Tuesday 28 June XIV 4. Nuku'alofa.

CASE-STUDY III

A multi-hazard history of Antigua*

THIS CASE-STUDY IS based on some of the annexes to a report to UNCTAD (made at the request of UNDRO, now the United Nations Office for the Coordination of Humanitarian Affairs: OCHA) on the social and economic effects of natural disasters in island developing countries (Lewis, 1982c). The report contains material on Antigua, Cape Verde, Comoros, the Cook Islands, the Maldive Islands and Western Samoa (UNCTAD, 1983).

Antigua experiences earthquakes, droughts and hurricanes. To isolate for study each of these as they occur would be to ignore the interrelationships between the after-effects of one and the effects of the next. Moreover, there will be conditions arising from factors outside the natural disaster spectrum which bear upon, and are themselves affected by, all of these phenomena.

This interplay of events and conditions is readily illustrated in the case of island countries, which have a natural and clearly defined containment. Such interrelationships suggest a complex human–ecological system which must be recognized if environmental balance and compatibility are to be maintained – particularly in respect of hazards. This documentary analysis of the colonial era in Antigua has to conclude for the time being with questions concerning the environmental effectiveness of superimposed systems of administration which, with incomplete knowledge of comparable natural hazards, assumed sectoral separation for their administration, as well as for everything else.

Introduction

Islands are an anachronism. Seemingly insignificant in global terms of physical size or population, they have often been of crucial strategic importance to world powers. Always heavily dependent upon trade but vulnerable to world economic and political fluctuations that they are unable to control, islands were early victims of protectionism and then colonialism. Able to contain their epidemics and sometimes to escape tropical cyclones that are caught by larger land masses, islands nevertheless sustain highest proportional social and economic damage when disasters are sustained. Whereas natural disasters were once an impediment or a bonus to colonial profits, they may now be the vehicle of favours for political advantage – a new disaster imperialism.

Disasters in history may be examined in a number of ways. First, disasters may have had implications for contemporary events so that history itself may have been directed by disasters (Stevenson, 1912); second, certain disasters may be

---

* Formerly published in *Disasters* 8:3, 1984.

examined in isolation, from records that exist of single catastrophic events (Hughes, 1983); third, disasters of similar kinds may be recorded from all sources as a global or regional review of, for example, earthquakes or storms (e.g. Eiby, 1957; Kerr, 1976); and fourth, disasters of all kinds may be assessed from available records concerning one location selected for study (Hewitt and Burton, 1971).

There are certain academic satisfactions accruing from the isolation of disasters from other subject matter and from the specific or general collection of data concerning them. Study of disasters of all kinds *in situ* to one place, however, more closely reflects the experience of natural hazards over time, the part they may have played in history and the part each has played in accumulating physical, social and institutional vulnerability to those that followed.

The degree and extent to which certain changes may have followed often disastrous failure (Davis, 1983) has to be balanced by systematic observation and analysis of how change may have been the cause or exacerbator of recurrent subsequent disaster of the same or a different type. In the history of the human ecological context that is thus revealed, response to one disaster may have exacerbated vulnerability to another. Form may have followed failure, but that did not necessarily stop the overall vulnerability process.

Certain globalized and general conclusions can be abstracted with regard to, for instance, earthquakes and other disasters upon building construction, but such a global collection would implicitly mean the abstraction of each earthquake from its local context into an amalgam of subject matter from which global conclusions concerning earthquake effects could be drawn, but which are then of only limited usefulness for their reintegration within specific contexts.

The environmental reality of places where earthquakes are experienced is that one earthquake precedes the next. Where one earthquake or tremor may have caused peripheral settlement, cracking, or other instability, the next may cause greater damage than if the first had not occurred. More significantly, earthquakes occurring in multi-hazard areas have a bearing upon, and their consequences will themselves be affected by, conditions created by disasters of all kinds.

The integration of disasters with other events in history will also reveal more of the specific reality of each social and natural environment. For example, one effect in March 1889 of hurricane upon the ships of three navies assembled in Apia Harbour (Western Samoa) was the subsequent negotiation for the Pacific in the Treaty of Berlin, instead of war (Stevenson, 1912); and the earthquake of 1843 in Antigua had a serious effect upon the emergent communities of then only recently emancipated slaves (Woodcock, 1843).

Abstraction of events from their historical contexts may have the misleading consequence of attributing a seriousness to earthquake or volcanic eruption that external observers might exaggerate. In their contexts, however, some events that would perhaps be serious elsewhere are not locally considered serious at all (e.g. eruption of Fogo, Cape Verde Islands: UNCTAD, 1983). Indigenous interpretations and perceptions of hazard must be the basis for their environmental evaluation and assessment. Drought in Antigua would not have been considered drought in the Cape Verde Islands, used to, and capable of, resisting far greater extremes (UNCTAD, 1983).

Drought in Antigua meant loss of sugar production and other exports, of serious consequence to colonial administrators. In this respect, however, a distinction has to be made between indigenous administrators and those in the corresponding metropolis, who were often in conflict – and often about how 'the natives' were to be treated after a disaster (UNCTAD, 1983).

The Earthquake of 1843

Earth tremors had been a common occurrence during the eighteenth (and early nineteenth?) century. On 16 May 1778:

> . . . the earth shook violently three or four times . . .many of the whites as well as negroes were much alarmed and ran out into the street' (Luffman, 1789).

But

> At 20 minutes before 11 o'clock on Wednesday morning the 8th February (1843) Antigua was visited by a dreadful earthquake. . . there arose clouds of dust from every part of the town, the crash of falling buildings was heard, blended with the piercing shrieks of the people and accompanied with that horrid heaving and trembling of the earth beneath our feet. . . Almost every piece of masonry in St John's is in ruins (Rev. H Cheeseborough: 10th February; Wesleyan Methodist Missionary Society (WMMS), 1843).

> The stone dwelling houses and stores were crashed and crushed . . . the wooden buildings waved to and fro . . . The damage done is immense. In the capital (St John's), some of the finest stores are a mess of ruins . . . and in many parts the earth is opened, forming deep fissures (Woodcock, 1843).

In St John's, the courthouse, police-office, arsenal, new jail, and barracks were:

> fearfully dilapidated. The Register Office, treasurer's office, Governor's Secretary's Office (just erected) and the Colonial Bank were all much injured.

All the stone buildings on Barbuda (except one schoolhouse) were destroyed. On Antigua at the dock-yard of English Harbour the

> wharves all rocked and rent; in some places they have sunk down to the margin of the sea, in others they are literally heaved up. . . (Woodcock, 1843).

Five stores built since the fire of 1841, and seven others, three taverns (one three-storey in brick); a brass and iron foundry ('the only one of its kind in the West Indies'), a bakery, private dwelling houses ('that is those built of stone or brick'), 'almost every kitchen and oven on the island' and cisterns, were destroyed or very severely damaged. All the 172 sugar mills and estates received damage, 35 were entirely destroyed, 82 irreparably damaged; 52 partially damaged; and 'works, dwelling houses, labourers' cottages attached to those mills shared their fate in equal proportions.' Numerous 'free-villages' built by their own labour by ex-slaves were destroyed (slavery was abolished by Great Britain by a law passed in 1834).

Many of the estates that have fallen prey to the earthquake have been established since emancipation, by men who have exerted themselves to the utmost. . . and how they will be able to rebuild them it is impossible to say. Indeed it will take many years to restore Antigua to its former position (Woodcock, 1843).

St. John's Cathedral was badly damaged and declared 'unfit for public service' and several parish churches were destroyed or badly damaged, as were eight chapels or mission houses, one 'not much, being a wooden structure'. The largest, the Eberneezer Chapel requiring £3000 to be rebuilt according to an estimate from HM Civil Engineer who advised:

To rebuild in stone would require less by about £500, and though the building would be liable to be damaged by earthquakes it would be less exposed to the ravages of fires and hurricane which are of more frequent occurrence (Keightly, 18th February: WMMS, 1843).

The capital, St John's, had been destroyed by fire in 1841, and it seems that much rebuilding had been completed in 'fire-proof' masonry. It is a source of contemporary comment that masonry buildings suffered most damage in the earthquake of 1843. Many houses were left with their outer masonry walls collapsed and with the inner wooden walls supporting the roof; houses built entirely of wood remained standing.

Nearly all our [Methodist] members in both town and country, are sufferers. . . some of them to an almost ruinous extent. Even the labourers, of whom a large proportion had invested the savings of eight years [since emancipation] of toil in the dwellings they had built have been reduced to such a state of destitution by the destruction of their tenements as to be literally homeless and penniless. . . (Keightly: WMMS, 1843).

There were various estimates of deaths, from 12 to 40, and the total cost of damage to the island, including the loss of the sugar crop, was placed at £2 million.

Setbacks to economic recovery

An Act was immediately passed requiring:

inhabitants to pull down all injured buildings, in order, if possible to guard against any further accidents. In case of neglect, a committee is appointed to do so, and £100 sterling granted to defray expenses, to be refunded by each individual, either in money or by sale of a part of the broken fragments (Woodcock, 1843).

A grant of £500 was placed at the disposal of the committee to support the cathedral roof, the restoration of some of the parish churches being commenced in 1845, the repair of those more seriously damaged having been completed with government funds by that time. A new cathedral was finally completed in 1846 at a cost to government expenditure of £35 000:

. . . heavy drain on the public resources; and the effects of this extravagance will, I fear, be sensibly felt for some time to come (Department of the Colonies (DC), 1847).

Methodists received nothing from public funds:

> . . .all is bustle and activity in the Establishment. The Legislative grants large sums of money for repair and rebuild . . . church after church rises from its ruins . . . (Keightly: WMMS, 1843).

In spite of increased expenditure for relief and reconstruction, the necessary increase in imported materials produced duty revenue for government funds. An excess of revenue over 'a very liberal expenditure' and a balance in hand at the end of 1845 of £13 717, 11 shillings and 10 pence* was recorded (DC, 1845):

> The increase in the actual receipts has arisen for the most part, from the augmented consumption of dutyable goods, and particularly the productions of the United States; although the declared value of imports generally was less in 1845 than the preceding year. [However,] the net excess of expenditure amounts to £8232 sterling, which has been caused, in great measures, by the unavoidable and heavy expense incurred in rebuilding the Cathedral and restoring other public buildings. . .

There is no record of how the decision was taken to rebuild the cathedral from public funds; but the cost of rebuilding was a source of irritation to HM Governor.

There was an accompanying decrease in the value of exports for 1845 of £107 530 indicating 'a considerable failure in the produce of island staples' (see Table 1).

The year 1846 saw a diminution in both imports and exports as compared with 1845:

> Falling off of imports appears to be chiefly attributable to a diminished quantity of supplies being introduced in the past year from the United States; arising partly perhaps from the more contracted demand for them than in previous years, when an unusual quantity of supplies of various kinds was required for the restoration of damages occasioned by the earthquake of 1843, and partly perhaps from the very short crop of 1846 causing money to be less freely circulated (DC, 1846).

The colonial report for 1847 is unusual in its inclusion of a detailed statement of accounts comparing 1847 with 1846. Significant increases in expenditure are shown for highways, purchase of land, and 'cost of iron tanks for Court-house' (rebuilding). There are decreases for 1847 shown, among other items, for forts and parishes, indicating perhaps higher expenditure in 1845 more closely following the earthquake. The largest item of decrease (£1940, 16 shillings and a halfpenny) is in fact against the item for 'Expenses from earthquake' with an aggregate expenditure (1846/1847) of £9791. Revenue accounts showed increases on almost all duties and licences, the marked decrease in tariff duties. 'Expenses

---

* 20 shillings = £1.00; 12 pence = 1 shilling.

Table 1 Antigua: Exports 1844 and 1845

		1844	1845	Deficit
Sugar	Hogsheads*	15,357	11,809	3,548
	Tierces	1,562	1,012	550
	Barrels	4,512	2,745	1,767
Molasses	Puncheons	9,020	8,780	240
	Hogsheads	127	–	127
Arrowroot	Boxes	665	407	258
	Barrels	104	–	104

(Source: DC, 1845)
* A hogshead was 15 hundredweight (average); three tierces = 2 hogsheads. 1 hogshead = 8 barrels. A puncheon was a large cask holding from 72 to 120 gallons.

of Earthquake' for 1847/1848 were £2060, and that year showed an even more marked falling off of post-earthquake reconstruction expenditure.

Parliament in London sanctioned an advance to Antigua in 1844 'towards remedying the destructive consequences of the earthquake in the preceding year'. At the end of 1854, the consequent public debt was £65 000 and:

the reductions which have been lately conceded by HM Government by the amount of the annual instalments of repayment of the principal, from one tenth, to one twentieth, and of the interest from a rate of 5 to one of 3 per centum, have rendered this obligation a comparatively light and easily man-ageable one (Governor MacKintosh: DC, 1855).

These concessions had been hard fought for (reading between the lines of colonial reports) and the obligation eased only temporarily. Governor Hamilton, in his report for 1856 wrote:

The heaviest liability under which the Colony suffers is the loan from Her Majesty's Government on the occasion of the calamitous earthquake of 1843. I do not now allude to the bulk of the amount lent, which was appropriated to the relief of the necessities of the individual sufferers, but to that portion of it which was retained for the public service, and was expended in the repairs of public buildings. . . the strain of this engagement is only now beginning to be felt.

The advance was made available in the form of loans by the Antiguan admin-istration to borrowers, who were due to repay by instalments to coincide with Antigua's ten yearly repayments to HM Treasury in London:

Had the petition to HM Government been for the remission of the portion which must be raised by taxation on a community only just recovering from the struggle of competition between free-labour and slave-grown sugar, their proceedings would at least have met with sympathy, even if they had not met with concurrence . . .

In 1860:

the debt to the Government has been reduced to £14 857 yet, as no separate provision has been made for the liquidation of any part of it, and as the ordinary income of the Colony was inadequate for that purpose, the means by which it has been reduced have been obtained by local loans, indicated by the debt due to the Savings Bank and issue of Treasury Notes. By the subsisting arrangement the debt to the Government is to be reduced in 1865 to £10 000 by the payment of annual instalments; and such £10 000 are being paid in moieties in the years 1866 and 1867' (DC, 1861).

The earthquake loan disappeared from colonial reports only in 1868. In 1867 construction commenced of a waterworks which continued for three years at a cost of £30 000, and a capacity of 500 000 gallons. This measure of attention to recurrent drought had had to wait until the burden of the earthquake loan had disappeared.

Drought

Throughout the period of colonial administration in Antigua the most important crop was sugar cane. Its success or failure in any year was the indicator of success or failure of the colony. Although colonial reports make reference in varying degrees to living conditions and other social factors, there is an overriding concern for income from sugar production, and the success of a governor's term of office was clearly dependent on revenue.

International fluctuations in the price of sugar itself had a much more serious impact than any other factor up to about 1900. Low prices often confounded high production, but in 1895, when very low prices accompanied very low production, it seemed that the sugar industry was doomed to extinction – saved only by a rise to average production in 1896 (Watts, 1906).

Before 1898, cane disease was the prevailing factor influencing production. It took many years of experience to distinguish the effects of disease and drought, but successful experiment with resistant cane brought disease under control by 1898. Thereafter the relationship between rainfall and sugar production became clear, though still masked in small degree by changes in agricultural methods, variations in acreage, new varieties of cane, and factory efficiency. Sugar production in the years immediately following 1900 was below average, due entirely to deficient rainfall and the damage caused by hurricane in 1899 (see below). Thereafter, the construction and equipping of centralized sugar factories and the introduction of mechanized ploughing and transportation, indicated a confidence in the future of the industry which, as it turned out, heralded a period of increasing annual average production.

The relationship between rainfall and sugar production was examined in a retrospective study of the 25 years 1930–1954, years of rainfall values being grouped and set against annual sugar production of the same years (Auchinleck, 1956) (Table 2).

The average rainfall for the longer period of 76 years (1874–1949) was lower, at 43.26 inches, than that in the Table. Years of rainfall significantly below this average were 1874, 1875, 1882, 1890, 1905, 1910, 1912, 1920, 1921, 1922, 1923, 1925, 1928, 1930, 1939 and 1947. In addition to these 16 years of severely low

Table 2 Antigua: Rainfall and sugar production 1930–1954

Rainfall of preceding year	Number of years	Tons of sugar: yearly average
Below 30 inches*	1	4,442
30–40 inches	4	15,626
40–50 inches	7	19,041
50–60 inches	9	20,010
60–70 inches	1	27,713
Above 70 inches	3	28,657
Average 50.88 inches	25	Average 19,760

(Source: Auchinleck, 1956)
* 1 inch = 254 mm.

Table 3 Antigua: Exports and imports 1871–1874

	1871	1872	1873	1874
Imports (£)	175,740	200,577	169,156	146,758
Exports (£)	247,630	53,190	170,977	106,705

(Source: DC, 1875)

rainfall, there were a further 17 years with rainfall below average. As Antiguan rainfall was gathered from a number of measurement stations, it is certain that some local conditions were worse, and some better, than the national averages. Over the same 76 years (1874–1949) there are, however, only 14 years where drought has been a significant claim in the Colonial Records. It can be accepted, therefore, that drought conditions, when officially reported as such, were economically and socially serious in the national experience.

Drought in 1863–1865 had an obvious impact on a mortality of 47.8 per 1000 population. A total of 5222 deaths were recorded for the period – 14.4 per cent of the population. The sugar crop of 1874 was the smallest since 1864, and the total value of all exports fell accordingly from £170 977 in 1873 to £106 705 in 1874. Related years are shown in Table 3.

Subsequently, at the end of 1912, Antigua had 'suffered from three successive years of drought, which caused considerable distress in country districts . . . The drought culminated in an almost complete failure of [water] supply in St John's, and for some days an acute water famine prevailed' (Colonial Office (CO), 1913).

The beneficial effect of hurricane in bringing rainfall and ending a three year period of serious drought was apparent in 1924; 'Hurricane brought damage of several thousand pounds but also brought relief in the form of welcome rains' (CO, 1925). Rainfall for the year was 41.57 inches, the heaviest on 27 August (preceding the hurricane of the 28/29 August), and was almost ten inches above that of the preceding year.

Hurricanes

As in the last case noted above, hurricanes on and near to Antigua have frequently brought beneficial rain, and to those engaged in sugar production the benefits of employment and income at all levels. Their immediate consequences have nevertheless sometimes been very serious, most significantly in 1681, 1772, 1780, 1792 and 1804, although a total of 22 have been collated for the period of 183 years 1664–1846 (Garriott, 1900).

The hurricane of 1848, though of serious impact, received scant mention in the colonial report for the year, which was still preoccupied with the aftermath of the 1843 earthquake. The hurricane of 8 September 1899 caused damage to houses, but no loss of life, though 'much damage to the huts of the labouring classes, who consequently suffered from exposure and distress' (CO, 1899). Its part in the run of poor years of sugar production after 1900 has been mentioned above.

The hurricane of 28–29 August 1924, which ended three years of serious drought, caused 'moderate' damage. A relief fund established by the Lord Mayor of London reached £4000 which was 'devoted to the relief of peasants and labourers and the reconstruction of their dwellings' in Nevis, Montserrat, Tortula, St Kitts, as well as Antigua, whose share was £1356, five shillings and ninepence. Of this amount, a sum of £500 (!) was placed on deposit 'as the nucleus of a fund to meet further similar disasters' (CO, 1925).

Contributions of clothing and food were sent from other West Indian colonies and England, the French West Indian colonies, the government of the Virgin Islands and the USA. The cost of reconstructing and repairing government property was met partly from a £10 000 grant from the London Parliament and from surplus funds (the total cost is not given). Total aggregate revenue for the year 1924–1925 was £78 983, eight shillings and ninepence, and total national expenditure was £85 244, 13 shillings and ninepence – a rare excess of expenditure over revenue (CO, 1925).

Following the hurricane of 1928, a special commission visited Antigua to assess and report upon hurricane damage (Collens, 1928). Under 'General Observations and Recommendations' their report stated:

1. Peasant houses. We have in all cases taken into consideration the age and condition of the houses at the time of the hurricane, and the ability or otherwise of the owner to meet the total or partial cost of repairs or rebuilding. The allocation of any hurricane funds for such destitute owners can in our opinion be left in the hands of the local authorities.

2. Damage to Government Buildings, Services, Telephone System, Press etc . . . [we] have differentiated between actual damage caused by hurricane effects and damage which may be attributed to normal wear and tear or natural causes . . . [and] have endeavoured . . . to apportion the estimated cost of renovation or renewals between Hurricane Relief Funds and the funds of the Presidency concerned . . .

(At this time [1928] Antigua was the principal seat of government in the Leeward Islands colony, which comprised the presidencies of Montserrat, Dominica, Nevis, St. Kitts and Antigua.)

3. In view of the well-known periodicity of hurricanes in these islands we would recommend that some general form be drawn up for universal use in each Presidency indicating the nature of damage, and its assessed value and the quantity of nails, lumber, boards, and shingles, if any, issued as relief or estimated as required for reconstruction.

Damage was assessed in categories: (a) for private houses (exclusive of estate property); (b) for private houses (requiring some possible assistance); (c) for private houses (poor and destitute persons) and (d) for government property. Total damage assessments for category (c) came to £2900; for category (d) to £2527, a sum £355 less than the local estimate. The Commission recommended special consideration for rebuilding the poor house at a cost of £2500 (extra to come from presidential funds):

as the Poor House is 28 years old having been hastily built to house Boer War prisoners, but never used for the purpose.

In 1927 the principal author of the 1928 report had amended a (then existing?) hurricane code (Collens, 1927) which focused principally on domestic precautions concerning shuttering for the prevailing wind, and warning symptoms of a falling barometer. 'Mutual telegrams (were to be) exchanged between islands of the Leeward Island Colony by the West Indian and Panama Telegraph Company.' A red flag with a square black centre would be hoisted as storm warning signal at Rat Island signal station. If a hurricane was to be definitely expected (or at night), 'two detonating rockets will be fired in rapid succession from the hill near the Botanic Station'. The 1928 Report does not comment on the efficacy of these measures of hurricane preparedness.

In 1950 there were two serious hurricanes (in addition to two serious fires) in St. John's (CO, 1950). The first, on 21 August, brought winds of up to 100 miles per hour and severe destruction in rural areas, deaths of livestock, and extensive local damage. Altogether, 488 houses were destroyed and 636 houses were damaged – 'many being rendered uninhabitable'. The second hurricane, ten days later on 31 August, brought 165 miles per hour winds and references to greater damage in the capital of St John's than in rural areas. There was considerable damage to government, private, and commercial dwellings and:

leaving out an account of large houses, which were either insured by their owners or whose owners could afford to repair them unaided, 1348 small houses were completely destroyed and 2343 damaged in both hurricanes.

In Antigua, 6477 people were made homeless. In Barbuda, an additional 84 houses were destroyed, 109 damaged, and 320 people made homeless. The total of 6797 homeless were 15 per cent of the total population of the colony.

His Majesty's Government (London) made a grant of £50 000 for relief, and the British West Indian Government made gifts of clothing, food and medical supplies. Jamaica gave £5 000. American and French territories also gave relief supplies. The homeless sheltered for many weeks in churches, schools and halls.

No further damaging hurricanes have been reported in the period up to and beyond the end of the colonial administration in 1967, although droughts recurred during this interval.

The Earthquake of 1974

There were no significant foreshocks for the earthquake of Richter magnitude 6.7 which occurred at 05.51 hours on 8 October 1974. That there were no deaths was attributed to the early hour of the event, when few people would have been about, and places of work, centres of congregation and commerce, and public buildings would have been unoccupied (Tomblin and Aspinall, 1975).

Severe damage was inflicted upon government buildings, the port, and infrastructural services of roads, electricity and telephones, and water supply. Government buildings severely damaged and rendered uninhabitable were Parliament, Judiciary, Treasury, Central Registry, two government ministries, the Secretariat of the East Caribbean Common Market, the Public Health Service complex, the library, printery and prison. The Anglican cathedral, rebuilt after the 1843 earthquake, received some significant damage; the prison was built in 1735 and had been severely damaged in the earthquake of 1843. The list of government buildings damaged in 1974 is very similar to those damaged in 1843, and the reasons much the same, all being unreinforced masonry or inadequately constructed reinforced concrete frame buildings. Half of the total accommodation being utilized for government operations were rendered unusable (ECLA, 1974).

The authorities of a country where drought is more frequent than earthquake were quick to make emergency repairs to damaged water mains and dams which reserved drinking water. Principal industrial damage was to the oil refinery, rupturing tanks and pipelines, causing a severe pollution hazard (and fire risk) and as the island's largest employer, the laying off of up to one-third of the workforce. The private sector suffered severely and an immediate scarcity of bread resulted from the destruction and damage caused to bakeries. Lobster reefs were damaged by the earthquake, with immediate commercial losses to the fisheries sector.

Three areas of concern were expressed for housing. First was with the 40 homeless households; second, with 800 habitable but damaged housing units where there was no insurance coverage and family earnings were too low to effect repairs without assistance; and third, concern with damaged housing with insurance cover inadequate to compensate the full cost of repairs. Housing losses were sustained mainly in the rural areas, and mostly to buildings of traditional construction inhabited by the lowest income earners (ECLA, 1974).

The 132 years that had elapsed since 1843 had made the Anglican and Catholic cathedrals, parish churches and chapels eligible for reconstruction assistance as Places of Historical and Cultural Interest as essential elements in the history of the country. The same period represents a significant interval of seismic quiescence, a quiescence of major earthquakes not ended by that of 1974, which was perhaps two orders of magnitude less than the 1843 event.

Postscript: The library remained closed due to damage to the building in 1974, and was still unrepaired in 1988 (Kincaid, 1988).

Conclusions

Colonial government was not a monolithic overburden. Disagreement and dispute between the local governor in Antigua and the Department of the Colonies in London over earthquake loan repayments continued for 25 years. Here the metropolitan authority was the most heavy-handed, but in the Comoro Islands after severe hurricanes in 1904 and 1905, it was the visiting officers representing metropolitan government in Paris who were less onerous in their assessments of compensation on behalf of the *indigènes* and who more heavily reassessed the *colons* (UNCTAD, 1983).

Issues that continue to preoccupy administrators of post-disaster assistance were being debated in the same terms in Antigua in 1928, as was hurricane preparedness. (Also in the South Pacific, 143 years ago, the impossibility of cooking post-hurricane rice, unknown in the Cook Islands, was a subject of polite but emphatic correspondence of the then incumbent missionary (UNCTAD, 1983).) The poorer resilience of masonry structures in respect of earthquakes, but their value against wind and fire were well known and consequently well observed in Antigua in 1843.

It is a simple matter to abstract these observations but it is the interrelationships of issues which is of predominant importance – and which a separation of studies would obscure. The effects of hurricane on a country still suffering from earthquake; the influence of fire upon building construction methods that increased vulnerability to earthquakes; the beneficial effects of a hurricane ending prevalent drought; delay to water storage programmes caused by the imperatives of earthquake loan repayment; and the effects of earthquake on the settlements of recently emancipated slaves, are all evident from these outline histories of hazards in Antigua.

This complex administration of human–ecological interrelationships in respect of hazards cannot logically be separated from the study of disasters. All things come together in islands to a degree not so easily discerned and identified in subcontinental or metropolitan countries. This characteristic is of crucial importance for management and development planning – in islands and at local levels of larger countries.

Systems of government adopted or inherited from former colonial powers assumed an administrative separation wholly inappropriate to both the place and the environment of islands. They also assumed the pre-eminence of economic factors on behalf of their colonies, which has been largely redressed by nationally elected governments. Now with different values, therefore, are inherited systems of administration the most appropriate to the management of hazards and other environmental issues, of which, after all, the former metropolitan powers had no direct experience of their own?

References

Auchinleck, G G (1956): *The Rainfall of Antigua and Barbuda. Compiled from available records* The Antigua Sugar Association, Antigua.

Collens, A E (1927): *Leeward Islands Hurricane Warnings and Amended Hurricane Code* Antigua.

Collens, A E (1928): *Leeward Islands; Report of the committee appointed to assess hurricane damage of september 11–12, 1928* Antigua.

Colonial Office (1886–1938; 1947–1954): *Report on the Blue Book (Leeward Islands)* HMSO London.

Davis, I (1983): Disasters as agents of change? or: Form follows failure *Habitat International* 7 (5/6) 277–310.

Department of the Colonies: *Reports on the Past and Present State of Her Majesty's Colonial Possessions*: For the years 1845–1848, 1849–1850, 1850–1851, 1852–1853, 1854–1855, 1856–1857, 1858–1859, 1860–1861, 1862–1863, 1864–1865, –1887.

Economic Commission for Latin America; Office for the Caribbean (ECLA) (1974): *Report on the Damage Caused in Antigua and Barbuda by the Earthquake of 8 October 1974 and its Repercussions* (Prepared by S St A Clark). ECLA/POS 74/15. 20th December.

Eiby, G A (1957): *Earthquakes* Frederick Muller. London.

Garriott, E B (1900): *West Indian Hurricanes* US Department of Agriculture. Washington.

Hughes, R (1983): Historic disasters *Disasters* 7(3) 161–163.

Kerr, I S (1976): *Tropical Storms and Hurricanes in the Southwest Pacific; November 1939– April 1969* New Zealand Meteorological Service. Wellington.

Kincaid, Jamaica (1988): *A Small Place* Virago Press. London.

Luffman, J (1789): *Brief Account of the Island of Antigua. Letters to a Friend: Written in the Years 1786, 1787, 1788* London.

Stevenson, R L (1912): *A Footnote to History: Eight years of trouble in Samoa* The Works of Robert Louis Stevenson, Vol. 17. Chatto and Windus. London.

Tomblin, J F and Aspinall W P (1975): Reconnaissance report of the Antigua, West Indies, Earthquake of 8th October 1974 *Bull. Seismol. Soc. Am.* 65 (6) 1553–1573. December.

UNCTAD (1983): *The Incidence of Natural Disasters in Island Developing Countries* TD/B/961.

Watts, F (1906): *Report on the Sugar Industry in Antigua and St Kitts-Nevis; 1881 to 1905* Colonial Reports – Miscellaneous. London.

Wesleyan Methodist Missionary Society (1833–1890): *Archives; Antigua.*

Woodcock (1843): *A Narrative of the Late Awful and Calamitous Earthquake in the West India Islands of Antigua, Montserrat, Nevis, St Christopher, Guadaloupe and etc on 8 February, 1843; Written by an Eye-Witness.* London.

CASE-STUDY IV

Vulnerability to a cyclone: Damage distribution in Sri Lanka*

INTERNATIONAL COMPARISONS OF natural disasters emphasize the significant vulnerability of poor countries and of the poorest people within the affected areas. Further comparisons suggest that small countries are likely to experience more serious disasters of national scale than large countries.

The research project described here sought to establish whether the approaches used and the patterns identified when comparing national-level disasters in various countries, have any relevance to the analysis of disaster within one country, using as a case-study, the impact of Tropical Cyclone 21 on Sri Lanka. Research set out to examine the relationship, if any, between socio-economic conditions and destroyed or damaged dwellings, as well as between population density and damage on the basis of information available in local records.

Tropical Cyclone No 21 in Sri Lanka

Tropical Cyclone No 21 originated as a tropical storm in the south of the Bay of Bengal on 17 November 1978 (Figure 1). Moving generally westwards, on 20 November it was classified as a severe cyclone, with sustained wind speeds of over 64 knots and, on 23 November at 5:30am, it crossed the eastern coastline of Sri Lanka at Batticaloa. Travelling northwest and moderating to a tropical storm, it continued across the Gulf of Mannar into southern India on 24 November and reduced further to a depression in the Arabian Sea (Meteorological Dept, Colombo 1979).

In Sri Lanka, sustained wind speeds were recorded at 80 knots (92 miles per hour) before the weather station at Batticaloa was destroyed, but based on satellite information the observatory at Colombo estimated the speeds to have been 125 miles per hour. Rainfall in the 24 hours (23–24 November) was recorded as 12 inches at Batticaloa. Heavier rainfall of up to 17 inches was recorded in the Central Highlands brought about by peripheral south-westerly winds of the cyclonic formation depositing rain on the high land. Severe flash flooding thus occurred in many places including Nuwara Eliya and at Ratnapura and Avissawella (Meteorological Dept, Colombo, 1979) (*Ceylon Daily News*, 1978) (Figure 2).

In all, probably two-thirds of the country was seriously affected, either directly by the cyclone or indirectly by severe flooding, or both. Reports from governmental and inter-governmental agencies and newspaper reports (e.g. *The Guardian*, 1978) inevitably underemphasized the total effects of

---

* Formerly published in *Ekistics* 308 September/October 1984 and in a longer form in *Marga* 6/2 1981. Marga Institute, Colombo.

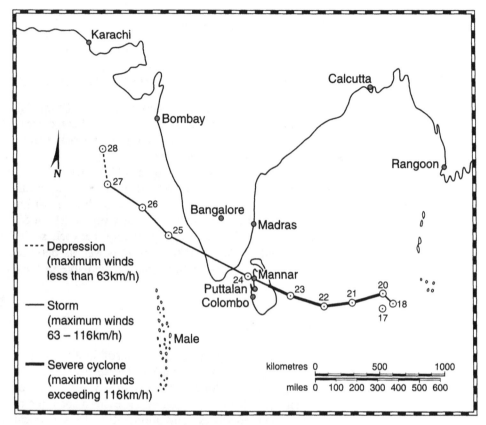

Figure 1 *Track of Tropical Cyclone 21, 17–28 November 1978* (Source: Sri Lanka Department of Meteorology in ESCAP, 1979)

disaster (Lewis, 1979a), focusing mainly on the urban centres, where perhaps physical damage was most obvious, and which were themselves the centres of communication.

Damage caused by the cyclone alone covered the whole of Batticaloa and Polonnaruwa districts and parts of Amparai, Anuradhapura and Matale districts, a total area of approximately 20 per cent of Sri Lanka (Figure 2) in which live approximately 7 per cent of the country's population (Department of Census and Statistics, 1978). Reported to be between 800 000 and one million (OFDA, 1978; UNDRO, 1979b), the numbers of people affected by cyclone and flooding were another 7 per cent of the national population, making the Sri Lanka cyclone disaster of 1978 closely comparable in these terms, with the Richter 6.4 earthquake which struck northern Greece in June 1978 (OFDA, 1978) and very much larger in national impact than the Yugoslavia earthquake of April 1979.

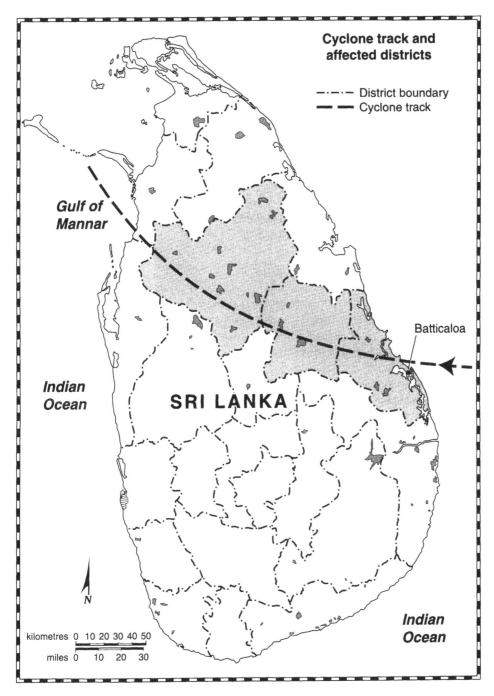

Figure 2

Damage analysis at the local level

Damage and the local administration

International disaster comparisons at the national level are important for under-standing the administrative and infrastructural burdens upon national govern-ments. Equally important, however, are comparisons of disaster impact at local administrative levels. The first administrative burden of disaster falls upon affected local governments. The burden of only localized disaster upon local government administration and local government area populations can be con-siderable (Lewis, 1979b). Local government administrations rely upon central governments when disaster recovery exceeds local resources. Where reliance upon central governments exceeds national capacity – as in the case of a large-scale disaster – the central government itself may become reliant upon external resources of disaster assistance.

Administrative structure of Sri Lanka

Sri Lanka is divided into 22 districts, each administered from a *kachcheri*, or secretariat, headed by a Government Agent (GA). Each district is subdivided into a number of divisions, directly administered by an Assistant Government Agent (AGA). Each AGA division is further subdivided into a number of smaller areas containing one or more villages, each represented by a *grama sevaka*, or village headman.

A significant feature of the 14 damaged divisions in four districts (Figure 3) is the much higher rural, compared to urban, population (Table 1). There are seven urban districts (municipal and town councils), with a total population of 108 122 (1971), but there are 992 villages with a total population of 438 963 (1977); and the density of rural populations, in divisions without urban areas, is in many cases higher than that of divisions that do have urban areas.

Damage by district

Total numbers of houses destroyed and houses damaged by division and district (Table 2) exceed, in all divisions, the figures for total housing stock by very much more than any margin of error, or increase in urban population since 1971. Figures of damage are in excess of totals for housing stock by an average over all divisions of 50 per cent. Figures concerning housing stock per population relate very closely to national average family size, and are therefore assumed to be accurate. Due to the lack of control in the collection process, figures on damaged and destroyed housing have been consistently overestimated or exaggerated. Although analysis is not reliable, the figure of destroyed housing has been used for the purposes of comparison with population density and socio-economic indicators.

Socio-economic indicators at the local level

Comparative studies of natural disasters in different countries have emphasized the significant vulnerability of poor countries (UNDRO, 1979a; Kates, 1979) and

90

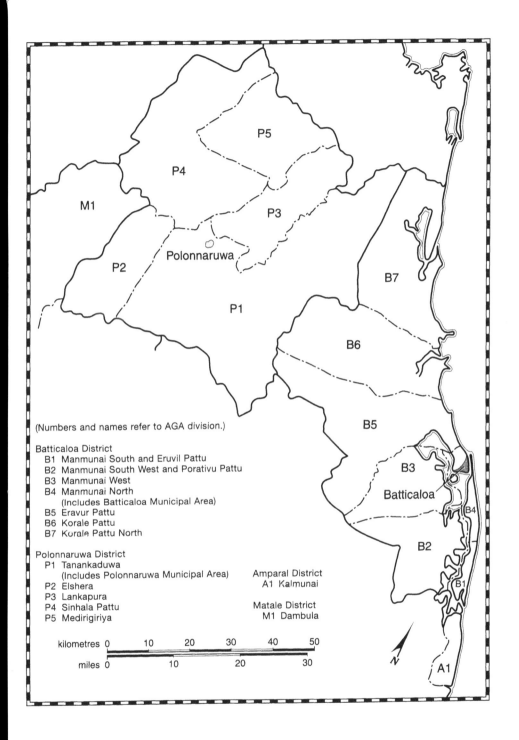

(Numbers and names refer to AGA division.)

Batticaloa District
  B1 Manmunai South and Eruvil Pattu
  B2 Manmunai South West and Porativu Pattu
  B3 Manmunai West
  B4 Manmunai North
       (Includes Batticaloa Municipal Area)
  B5 Eravur Pattu
  B6 Korale Pattu
  B7 Korale Pattu North

Polonnaruwa District
  P1 Tanankaduwa
       (Includes Polonnaruwa Municipal Area)
  P2 Elshera
  P3 Lankapura
  P4 Sinhala Pattu
  P5 Medirigiriya

Amparal District
  A1 Kalmunai

Matale District
  M1 Dambula

Figure 3 *The AGA Divisions of the most affected Districts of Batticaloa and Polonnaruwa; and Dambula and Kalmunai of Amparai and Matale Districts.*

91

Table 1 Area population, settlement and densities by damaged district and division

District / Division	No of urban areas (a)	Total urban population (1971) (a)	No of villages	Total rural population (1977) (b) (a)	Area in sq mls	Rural population density (p/sq ml)	Total pop (urban/rural)	Overall density (p/per sq ml)
Batticaloa								
B1	–	–	26	33,984	17.42	1,950.9	–	
B2	–	–	102	38,315	112.5	340.6	–	
B3	–	–	109	13,306	107.26	124.1	–	
B4	2	52,549	36	24,963	27.4	911.1	77,512	2,829.0
B5	1	16,959	95	30,432	227.125	134.0	47,391	208.0
B6	–	–	185	54,979	298.0	184.5	–	
B7	–	–	93	9,295	161.875	57.4	–	
Polonarua								
P1	1	9,684	82	38,266	580.29	66.0	47,950	82.0
P2	–	–	31	21,611	141.27	153.0	–	
P3	–	–	35	32,690	151.25	216.0	–	
P4	1	6,603	57	35,984 }	449.0	163.6	42,587 }	178.0
P5	–	–	52	37,464 }			– }	
Amparai								
A1	1	19,180(c)	16	37,511	8.5(d)	4,413.1	6,691	6,679.0
Matale								
M1	1	3,147	73	30,163	169.0(e)	178.5	33,310	197.0
Totals		108,122	992	438,963	2,450.89	–	305,441	–

Sources and notes:

a: Department of Census and Statistics, Colombo
b: Department of Census and Statistics, Colombo
Basic village-level statistics 1977
c: Source as a: above. In 1971 Kalmunai was a Town Council area before being made an AGA Division. Necessarily, reference to 1971 figures may reflect a higher urban value than the present actual urban area with Kalmunai AGA Division
d: Government Agent Kachcheri, Amparai District
e: Government Agent Kachcheri, Matale District

Table 2 Destroyed and damaged housing units by district and division

District	Division	Area (sq mls)	Total number housing units	Housing density (housing units/ sq ml)	Urban housing units (1971) (c)	Rural housing units (1978) (d)	Housing units destroyed		Housing units damaged	
							Number of houses (a)	% total housing units	Total housing units	% total housing units
Batticaloa										
	B1	17.42	6,019	345.5	–	6,019	4,812	80.0	2,768	46.0
	B2	112.5	6,928	61.6	–	6,928	3,347	48.3	7,444	100 (?)
	B3	107.26	2,282	21.3	–	2,282	2,605	100 (?)	834	36.5
	B4	27.4	10,779	393.4	6,233	4,546	5,128	47.6	11,905	100 (?)
	B5	227.125	11,737	51.7	7,265	4,472	9,983	85.1	4,033	34.4
	B6	298.0	10,412	34.0	–	10,412	8,744	84.0	3,719	35.7
	B7	161.875	1,809	11.2	–	1,809	1,792	99.0	615	34.0
Polonarua										
	P1	580.29	8,802	15.2	1,516	7,286	12,887	100 (?)	6,625	75.3
	P2	141.27	3,610	25.6	–	3,610	867	24.0	3,687	100 (?)
	P3	151.25	4,874	32.2	–	4,874	5,150	100 (?)	2,322	47.6
	P4	449.0	7,038	30.2	1,273	5,765	3,000	42.6	6,240	88.6
	P5		6,536		–	6,536	863	13.2	4,500	68.8
Amparai										
	A1	8.5 (b)	10,332 (f)	1,215.5	3,473 (f)	6,859	465 (b)	4.5	12,015 (b)	100 (?)
Matale										
	M1	169.0 (c)	6,048	35.8	610	5,438	1,064 (c)	17.6	5,903 (c)	97.6

Sources and notes:
a: Government Agent Kachcheris for Batticaloa and Polonarua
b: Government Agent Kachcheri, Amparai District
c: Government Agent Kachcheri, Matale District
d: Department of Census and Statistics, Colombo Basic-village level statistics, 1977
e: Department of Census and Statistics, Colombo
f: Source as e: above. In 1971, Kalmunai was a Town Council Area before being made an AGA Division
?: Cases where figures exceed census figures for housing stock

93

stressed that small countries are likely to suffer disasters of greater impact on the national scale than large countries (Lewis, 1979a). An early paper (Baird *et al.*, 1975) described the process as a result of which the poor become the most vulnerable to disaster and how, in disaster, the poorest are invariably identified as having been most vulnerable.

As a result of retrospective disaster analysis, both in theory and in the field, it may be useful to examine possibilities for relative field quantification and identification of socio-economic indicators so that sectors of highest vulnerability can be specifically incorporated into strategies for preparedness and preventive development planning.

Monetary indicators

International comparisons of gross national product and gross national product per head are approximate guidelines of relative welfare between countries. There are problems concerning international exchange rates, measurement of goods and services that do not enter international trade, arbitrary valuation of production for subsistence – usually a large proportion of total production in less developed countries – and valuation of production geared to varying needs within each country, all of which contribute to a final approximation of national production. Although GNP per head may be the best single overall indicator of differences in standard of living among countries, it is necessary for it to be supplemented by other indicators (Elkan, 1976).

Non-monetary indicators of relative welfare between countries are, for example, average life expectancy, energy consumption per head, average literacy, and calorie intake per head. There have been some attempts to overcome the problem of aggregating non-monetary indicators in order to find a more satisfactory alternative expression for comparison of the level of living (Grant, 1979; Hicks and Streeten, 1979). One significant selection uses steel consumption, cement production, the number of letters sent, the national stock of radio receivers, telephones and motor vehicles, and the consumption of meat (Beckerman, 1966).

The measurement and comparison of socio-economic indicators of different areas within a country cannot follow the pattern of measurement and comparison among countries at the national level, although it will be based on the same principles. At local levels:

- monetary indicators are further removed from reality, even if they are available. Money amounts may be minimal and the valuation of production arbitrary. Moreover, many
- nonmonetary indicators, found to be satisfactory at national level, may not be appropriate or may not be statistically available at local levels. Obvious examples are those of industrial production. Others – for example, numbers of motor vehicles and numbers of letters sent – while theoretically providing a basis for local comparisons, are available in statistical form only at national level, or for large regions, inappropriate for local comparison.

Recognizing these shortcomings, the United Nations Research Institute for Social Development (UNRISD) devised an alternative method to measure real

94

development progress at the local level in developing countries (Scott *et al.*, 1973). In attempting to formulate new indicators for the measurement of progress at local levels the authors stated that some of the data are notoriously difficult to collect – for example, personal income distribution are among 'the best guarded secrets' – such as the numerical strength of social groups and the size of land holdings. The authors acknowledged that 'the significance of an indicator can depend very much upon the local context', but the report also stated that 'At the village level. . . one can find real data, without extensive income and expenditure surveys, which particularly describe the participation in development of the lower castes or classes'. Most of the data selected for indicators can be interpreted as local socio-economic capacity and resources, the very same resources necessary for local preparedness and preventive development measures (Lewis, 1975; Lewis, 1978), in addition to their usefulness in measuring change in local progress.

In seeking to examine any relationship that there may be between local socio-economic indicators and the distribution of damage in post-disaster conditions, retrospective analysis of disaster will require information of the pre-disaster condition. Any method of assessment of progress which depends upon observation, such as the UNRISD method, is unlikely to be applicable in retrospect, and will not work at all where the very elements for observation have been destroyed, damaged, dislocated or made inaccessible by the disaster itself. On the other hand, a local survey for prospective analysis of future disaster damage requires the pre-selection of a location in which disaster is likely to occur in the near future, which is questionable in itself as well as being impracticable for short-term research purposes. It may, however, be applicable locally by indigenous national and local governmental authorities in cases where it is necessary to identify precise areas of particular vulnerability which are in need of allocated resources for preparedness and preventive strategy.

Research method adopted in the case of Sri Lanka

Retrospective, post-disaster research in Sri Lanka, six months after the cyclone of November 1978, with the purpose of assessing the relationship between the distribution of disaster damage and local socio-economic indicators required, first, identification of the local unit of area administration at which both data in socio-economic indices and disaster damage were available.

Because parts of districts were affected and because only two whole districts were involved, data at district level were insufficient for any realistic comparison to be made. Records of disaster damage had been made in some detail in respect of districts for the purpose of reporting to central government in Colombo (Government of Sri Lanka, 1978). These records had been prepared on the basis of information gathered by the *grama sevakas* and AGAs in turn, and collated by the GA in each district *kachcheri*. The smallest unit at which data on disaster damage were locally available in written form was the AGA division (Tables 1 and 2).

The most significant statistics available for use as socio-economic indicators for AGA divisions are the result of the Basic Village Statistics Survey made in 1977 (Department of Census and Statistics, Colombo, 1979). These surveys give figures for:

- population by age and sex
- population by major occupation
- unemployed population by educational status and sex
- distribution of housing units, households, number of villages with electricity and total number of villages
- land utilization (figures not available)
- distribution of families by land ownership
- numbers of fishing craft owned by villagers
- distribution of livestock and poultry
- distribution of industries (large-, medium- and small-scale)
- number of institutions by type of cottage industry
- number of housing units by type of cottage industry.

Attempts to determine other indicators applicable to, or compatible with, AGA divisions were unproductive in most cases. The figures of registration of motor vehicles and tractors were available by districts only; figures for the registration of radios, if reliable at all, were collated via post offices, the location of which is not related to GA or AGA administration – and registrations could be made at a post office in one area by persons residing in another; all health statistics are based on health administration areas, which are different from AGA district or divisional areas and, although considerable, were not therefore compatible with information on disaster damage.

It is useful to record an apparently direct relationship between socio-economic measure and the incidence of malaria, for which statistics are normally very detailed as a pursuit of anti-malaria programmes. Further research and analysis are necessary to determine how far data in the distribution of malaria incidence could be used as a socio-economic indicator itself (Ruberu, 1976; Visvalingam *et al.*, 1972).

Some figures were available on costs of agricultural production and farmers' incomes but for districts only, and other figures which had been prepared were not relevant to the areas affected by the cyclone. There are no official figures on the stock of telephones by division, but information on the number of telephones per AGA division has been abstracted from the telephone directory (Sri Lanka Post Office, 1978). This is the only information used at national level as a non-monetary indicator of wealth that is usable at AGA divisional local level (United Nations, 1977).

From village-level survey statistics, available per AGA division, some were not of use or were inappropriate as indicators of wealth. Population figures, distribution of housing units and total number of villages were basic data; households per housing unit were unreliable as an indicator of wealth without more information on social norms; land utilization figures were not available; number of fishing craft were not representative overall, being more significant to coastal areas; distribution of industry itself was considered of little value without additional information on numbers of people employed and from which AGA division, and similarly with cottage industry institutions. Number of pigs and number of cows per head were omitted, being unrepresentative due to cultural constraints, but numbers of poultry, goats and water buffalo were used – the figure for 'tractors in use' is very low (United Nations, 1977) and ownership of water buffalo is therefore significant as an indicator of wealth.

The indicators finally selected were:

1. Percentage of unemployed per employable population, ie: 15 to 54 year plus over 55 year age-groups.
2. Percentage of landless families.
3. Percentage of landless families plus families with less than half an acre.
4. Number of poultry per head.
5. Number of goats per head.
6. Number of water buffalo per head.
7. Percentage of villages with electricity (indicating an overall capacity per village to pay for electricity).
8. Number of telephones per division (indicating an individual ability to pay).

The selected indicators and their corresponding values for Batticaloa District rural areas are given in Table 3.

As indicators do not have equal significance, some weighting is necessary before indicator values per area can be computed to a single indicator factor. To find a method of weighting that is other than arbitrary has caused economists considerable concern (Beckerman, 1966). Beckerman's method is to find which national non-monetary indicators are highly correlated with aggregated national accounts. This method obviously cannot be made to apply for areas smaller than the national, but it is possible to determine which indicator most closely represents all the others for each division.

A computer program to determine the correlation matrix and therefore the most representative indicator has been applied to the values in Table 3 (Hall, n/d). Correlation output and correlation matrix are given in Table 4. The highest correlation between the eight selected indicators are the percentage of villages with electricity (EL) and the number of telephones per division (TE), but even these are not significant at the 0.05 level. The percentage of rural dwellings connected to electricity is very low, at 2 per cent in 1970 (World Bank, 1976). The percentage of villages served can therefore be assumed to be very low, and discounted. The distribution of telephones by individual dwelling or business is in any case a more accurate indicator.

The twelve divisions within Batticaloa and Polonnaruwa Districts, crossed by or closely adjacent to the track of the cyclone, have been used as the basis for analysis and comparison, with the addition of two adjacent divisions: Kalmunai in the coastal Amparai District and Dambulla in the inland Matale District (Figures 2 and 3).

Figures for each AGA division obtainable from the district GA *kachcheris* were for numbers of deaths, numbers of houses destroyed and numbers of houses

Table 3 Selected indicators: Batticaloa District rural areas

	1	2	3	4	5	6	7	8
	% unemployed	% landless	% landless + < half acre	Number of poultry per head	Number of goats per head	Number of w buffalo per head	% villages with electricity	Number of telephones per division
Division	UE	LA	LH	PO	GO	BU	EL	TE
B1	17.64	14.20	81.98	0.85	0.13	0.11	26.92	16
B2	4.50	22.42	30.38	0.69	0.31	0.44	2.94	4
B3	24.08	32.63	43.41	1.36	0.99	0.38	0.92	1
B4	14.97	13.22	85.46	1.72	0.16	0.002	30.95	52
B5	12.95	22.36	56.67	1.17	0.16	0.19	9.47	34
B6	7.38	22.93	74.05	1.43	0.50	0.34	4.86	27
B7	19.96	41.06	62.65	1.29	0.45	0.20	1.08	1

Table 4 Correlation output and correlation matrix

Correlation output

Selected social indicator		Mean	Standard deviation
% unemployed	UE	14.4971	6.8892
% landless	LA	24.2600	9.8530
% landless + less than half acre	LH	62.0856	20.2853
No of poultry per head	PO	1.2157	0.3509
No of goats per head	GO	0.3857	0.3042
No of water buffalo per head	BU	0.2374	0.1566
% villages with electricity	EL	10.9628	12.5210
Number of telephones	TE	19.2857	19.4054

Correlation matrix

Col / Row	UE 1	LA 2	LH 3	PO 4	GO 5	BU 6	EL 7	TE 8
UE	1.0000	0.3900	0.1644	0.3226	0.4393	-0.3138	0.0679	-0.2429
LA	0.3900	1.0000	-0.4506	0.0748	0.6523	0.4459	-0.8204	-0.6893
LH	0.1644	-0.4506	1.0000	0.4996	-0.4489	-0.8215	0.7525	0.6468
PC	0.3226	0.0748	0.4996	1.0000	0.2284	-0.4145	0.1835	0.5561
GO	0.4393	0.6523	-0.4489	0.2284	1.0000	0.6175	-0.6545	-0.5660
BU	-0.3138	0.4459	-0.8215	-0.4145	0.6175	1.0000	-0.8402	-0.6695
EL	0.0679	-0.8204	0.7525	0.1835	-0.6545	-0.8402	1.0000	0.6903
TE	-0.2429	-0.6893	0.6468	0.5561	-0.5660	-0.6695	0.6903	1.0000

damaged (in some cases divided between masonry houses and non-masonry houses) and, for Batticaloa District only, numbers of destroyed and damaged weaving centres and handlooms. Numbers of houses destroyed and numbers of houses damaged were the two categories that were common to all areas.

Damage was, of course, widespread sectorally as well as geographically. Beside damage to housing, there was damage to agriculture, fishing, industry and infrastructure. The number of houses destroyed is selected from the available data, not only as a significant aspect of sectoral damage in itself, but also as one that is representative of all divisions and districts, and therefore can itself be taken as an indicator of overall damage sustained.

Housing unit density

At first glance, Figures 4 and 5 show little positive relationship between the number, or the percentage, of housing units destroyed, and housing unit density (number of housing units per square mile); in fact, there is a tendency towards negative relationship. Both graphs relate the highest density A1 with the lowest number of units destroyed and B4, the next highest density related to the next lowest number, and percentage of housing units destroyed. Both these divisions contain very high population and are characterized by very high housing unit densities, and both contain urban or municipal areas (Table 1).

When all the divisions containing urban areas are taken on their own (identi-fied on the graphs) there is a strong negative relationship between density and

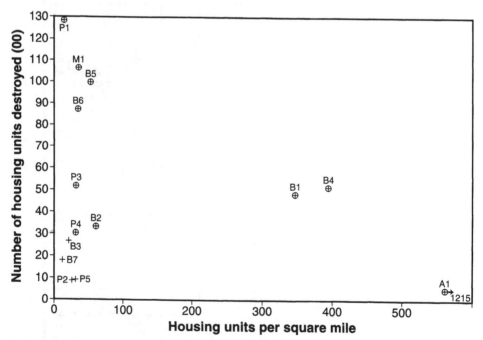

Figure 4 *Number of housing units detroyed and housing unit density (⊕ = Divisions with urban areas or high density)*

100

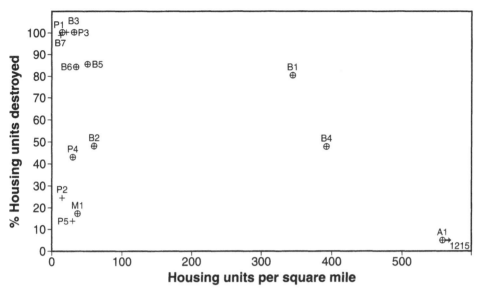

Figure 5 *Percentage of housing units destroyed and housing unit density (⊕ = Divisions with urban areas or high density)*

destruction. There is highest destruction in the lowest densities, P1, M1, B5; and lowest destruction in the highest densities, A1 and B4. On the same line of negative value, B1 is a high density non-urban division on the densely populated coastal belt, between A1 and B4 (Figure 3). The population density of B1 is higher than that of four divisions containing urban areas. Finally, B2 and P3, also high density, non-urban areas, are outside this negative value line, but these two are in the same relationship with each other. Clearly, P3, with highest destruction, has lower density and B2 with lowest destruction has the higher density.

As the cyclone moved inland it moderated (Figure 1), and wind speed reduced. The greatest wind speeds were therefore at the coast where, in this region of Sri Lanka, most concentrations of population are located. Of the three coastal divisions of highest population density, A1 and B4, show only medium to low percentages of houses destroyed. Of the remainder, the higher percentages of houses destroyed (given as 100 per cent or near) are in the low density divisions of B3, B7, P1, P3, B5 and B6.

Lowest density divisions have low percentages of destruction in B2, P4, P2, M1 and P5. Of this last low density group, the environmental factor of cyclone intensity appears to supersede the density factor. All the divisions are inland; B2 with the highest percentage is nearest the coast. On the other hand, the significance of A1 with very high density and low percentage houses destroyed, is due less to its distance from the high intensity centre of the cyclone track. In the divisions of very high density and very low density are the levels of lowest housing destruction, both in number and in percentage. Highest numbers of housing units destroyed are in divisions of intermediate density M1, B5 and B6 (Table 2).

Damage and destruction to housing units are lowest, as might be expected, in low density areas. What is more surprising, however, is that the highest density areas also offer some considerable protection. However, in this cyclone, the highest density areas are the coastal belt which would have received the highest cyclone intensity. Protection afforded by highest density supersedes the environmental factor of cyclone intensity, whereas the 'protection' against high number and percentage of destruction afforded by low density areas does not. Intensity overcomes low density 'protection'.

It may be regarded as significant that divisions P1, M1, B5, B6 and also P3 and P4, which have the highest, or a high, number of housing units destroyed are on the line of highest cyclone wind speed intensity. The cyclone inland would be comparable to the cyclone wind speed experienced in the divisions in the periphery as the coastline was crossed, i.e. the intensity at P1 would approximate to the intensity at B7 and B3. There are not sufficient wind speed measurement points to provide enough data to draw isopleths for hurricane intensity. Of the few anemometers, some were destroyed by the hurricane before it reached its peak intensity; observed damage being one method used to establish hurricane strength! The presence of P1, M1 and B5 at the top of the destruction scale (Figure 4) is therefore not due, or not only due, to hurricane intensity.

Socio-economic indicator TE (number of telephones)

The comparison of TE with number of housing units destroyed (Figure 6); appears to show conflicting positive and negative relationships until a separation

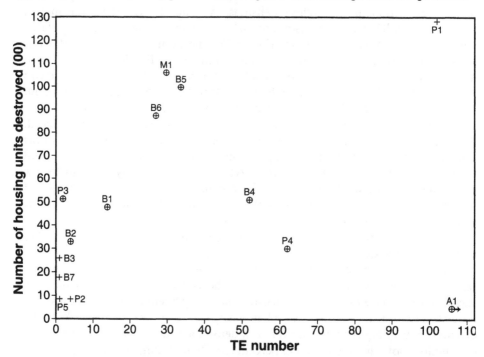

Figure 6 *Number of telephones (as a socio-economic indicator) and number of housing units detroyed (⊕ = Divisions with urban areas or high density)*

of values is made. From the cluster of six low density/low TE divisions there is an indication by four more of a definite positive relationship. The higher the TE indicator, the higher the number of housing units destroyed. To some extent this can be simply explained by the less definite but totally positive relationship between TE and total number of housing units (Figure 8): but the top values of positive relationship, M1 and B5, are the two lowest density divisions containing urban areas, after which all remaining values are similarly of urban content, showing a negative relationship. In this later group, the highest socio-economic indicator has lowest housing destruction. Low density areas show an increase of housing destruction with an increase in TE; high density areas show a decrease of housing destruction to increasing TE. The divisions of highest destruction are in divisions of intermediate TE value, as in Figures 4 and 5. The isolation of P1 is inexplicable.

The environmental factor of cyclone intensity may be significant in the relationship of the cluster of six low-density divisions (Figure 2), but ceases to be significant beyond. The environmental factor may predominate in low TE, low damage divisions. The correlation of TE and percentage housing units destroyed (Figure 7) shows neither positive nor negative definition.

In Figure 8, TE relates positively to total number of housing units. Divisions with highest numbers of housing units have the higher socio-economic indicator. Housing unit density itself, therefore, with more definite interpretation of relationships with housing unit destruction, becomes the more obvious key indicator, either for low- or high-density divisions. Nevertheless, the reasons why high-density divisions suffer lower percentages of destruction can be partly

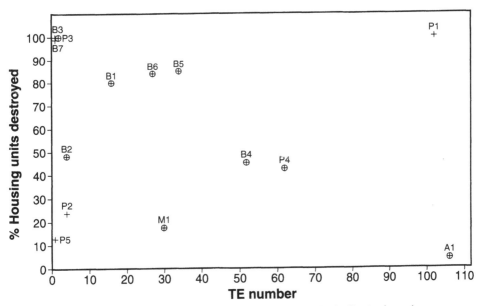

Figure 7 *Number of telephones (as a socio-economic indicator) and percentage of housing units detroyed (⊕ = Divisions with urban areas or high density)*

103

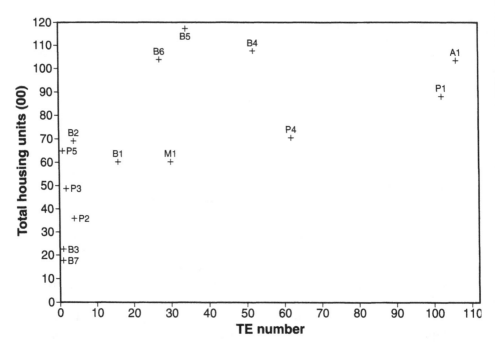

Figure 8 *Number of telephones (as a socio-economic indicator) and total number of housing units detroyed (⊕ = Divisions with urban areas or high density)*

explained in terms of higher quality of construction, that is, in terms of its socio-economic measure. Buildings also protect each other. The close compatibility between housing unit density and socio-economic indicator can be simply explained, but not explained away.

Examination of the relationship between divisional area and percentage of housing units destroyed (Figure 9) shows two groups of values of negative relationship. Each group shows a higher percentage of housing destruction for divisions of lowest area. The two groups are not, in this case, separable according to density, each group containing areas of high, and of low density. The group of highest percentage of housing unit destruction does, however, contain five of the divisions of highest cyclone intensity. If this grouping can be accepted as significant, there is a close relationship between a high percentage of housing unit destruction and small divisional area. The smallest areas suffer higher percentages of destruction.

Deaths caused by the cyclone (Figure 10) were available by division, only in the districts of Batticaloa, Amparai and Matale. Again, among the highest density divisions of B1, B2, B4 and A1, there is a definite negative relationship. The higher the population density, the lower is the loss of life. There is no definite relationship, negative or positive, between deaths and TE.

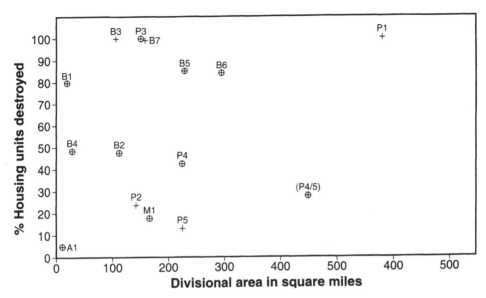

Figure 9 *Divisional area and percentage of housing units destroyed (⊕ = Divisions with urban areas or high density)*

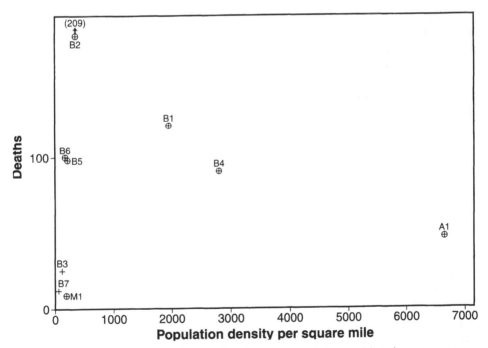

Figure 10 *Deaths and population density (⊕ = Divisions with urban areas or high density)*

Conclusion

Conclusions, or suggestions, are conditioned by the overriding factor of data incompatibility. The totals of housing units destroyed and housing units damaged exceed the figures for housing unit stock (Table 2). Inaccuracy in the gathering of statistics for disaster damage in disrupted working conditions, or intentional exaggeration of disaster damage for political or practical purposes, have been noted in other contexts. That either factor is present here is a small contribution to the stock of arguments for the impossibility of achieving an adequate database for post-disaster assessments not based solely upon observation. Further, some of the maps required for the identification of divisional boundaries, subject to considerable and recent readjustment and displayed on walls of administrators' offices, had been destroyed by rain after destruction of the roofs of the administration buildings. It is in the nature of the subject that post-disaster field research is likely to be disrupted, and results qualified accordingly.

Conclusions and suggestions offered from these analyses are summarized as follows. Although the strongest definition of relationship is between the socio-economic indicator (TE) and number of housing units destroyed (Figure 4), it can be argued that, because of the obvious positive relationship between TE and housing unit number in a given area (Figure 6), housing unit density is itself, in this case, the key indicator. TE is valid as a socio-economic indicator but is not sufficiently representative, and has been swamped by the more significant housing unit density. However, the two are obviously compatible, and if problems in data collection and reliability in another study could be overcome, it would be worthwhile to make another attempt to apply socio-economic indicators.

Number and percentage of housing units destroyed increase with housing unit density in lower density areas, but decrease in higher density areas, even though the high density areas are coastal and where cyclone intensity is highest. Hitherto, it has usually been assumed that areas of highest population density will suffer greatest losses, that is, that high density areas have the greatest vulnerability. These analyses now show that low density, that is (usually) rural areas, contain the highest number of destroyed housing units. The relative low number of destroyed housing units in high density areas is not always explained by those areas all being of high socio-economic measure (TE); nor by their location in areas where cyclone intensity was lowest. Not all high-density areas are of high socio-economic indicator value, and the highest density areas are coastal. Buildings protect each other. Additionally, the highest percentages of destroyed housing units are in the divisions of smallest area.

The obviousness of damage to buildings in a concentrated urban community and relative ease of access for reporters, assessors and administrators to and from urban centres, should not be permitted to obscure the possibly much greater social significance of damage to dwellings and other buildings and infrastructure in non-urban areas, which may not be so easily accessible but which may contain very much larger numbers of people.

Specific measures will be required to reduce, overall, the high incidence of

106

housing destruction and damage, as part of disaster prevention strategy, but in the direction of resources for disaster preparedness and preventive development planning. Whereas urban centres will also be centres for administration, transport, supplies and communications, these and all other services must be provided on behalf of, and with the participation of, the rural areas as well, for measures to correspond realistically to areas of need. These measures should additionally recognize the greater scale of potential need in the smaller administrative areas.

Notes and acknowledgements

The fieldwork for this study was undertaken during May and June 1979. The author's most sincere thanks are due to the director and staff of the Department of Census and Statistics, Colombo, for their kindness and co-operation, and in making pre-publication village-level census data available; also to the government agent and Kachcheri staff in the Districts of Batticaloa, Polonnaruwa, Anuradhapura, Amparai and Matale for hospitably attending to requests for information, both in person and by correspondence; to the Department of Meteorology for detailed information on the cyclone itself; and to personnel in many other ministries and departments, in Colombo and in the cyclone-affected districts and divisions, too numerous to mention separately. He is further grateful to Mr K W Tilakaratne for his painstaking abstracts from the Sri Lanka telephone directory. Most particularly, the author is grateful to Edward Horesh of the Institute for International Policy Analysis, University of Bath, for his advice during the preparation of this paper, and for his comments on a draft version.

References

Beckerman, W (1966): *International Comparisons of Real Incomes* Development Centre, Organization for Economic Cooperation and Development. Paris.

Ceylon Daily News (1978): November 25. Colombo.

Department of Census and Statistics (1978): *Statistical Pocket Book of the Democratic Socialist Republic of Sri Lanka*. Colombo.

—— (1979): *Basic Village Statistics 1977, Batticaloa District, Polonnaruwa District* (pre-publication version). Colombo.

Elkan, W (1976): *An Introduction to Development Economics* Penguin Economic Texts. Harmondsworth.

ESCAP (1979): Damage caused by tropical cyclones, floods and droughts in individual countries or areas in the ESCAP region 1978: Sri Lanka. Economic and Social Commission for Asia and the Pacific *Water Resources Journal* UN Bangkok.

Government of Sri Lanka (1978): *Parliamentary Debates (Hansard) Official Report* (uncorrected) Vol 3(1) No 7 Thursday 7 December. Colombo.

Grant, J P (1979): A new speedometer to track social progress *International Development Review* Vol 23 No 1.

The Guardian (1978): 25, 27, 30 November and 2 December. London.

Hall, E: *Time Series Processor*: Fortran IV version for the CDC 6600. Originally written for the CD6 6400 by Robert E. Hall, Department of Economics, University of California, adapted for the 7094 at the Harvard Computing Center by Frank C. Ripley and John Brode (program also available on the Harvard IBM 260/50, adapted for the CDC

6600 at London University by Sherman Robinson LSE). Processed at the University of Bath Computer Unit.

Hicks, N and Streeten, P (1979): Indicators of development: The search for a basic needs yardstick *World Development* Vol 7 pp 567–80.

Kates, R W (1979): *Climate and Society: Lessons from recent events* World Climate Conference, Geneva. World Meteorological Organization.

Lewis, J. (1975): Proposals for a working method of indigenous resource coordination as part of a predisaster plan *Occasional Paper No 3* Disaster Research Unit, University of Bradford. January.

—— (1978b): Comprehensive analysis of vulnerability to natural disaster; The socio-economic component Mimeo. 26 pp. February.

Meteorological Department (1979): Unpublished Data. Colombo.

OFDA/AID (1978): *Sri Lanka-Cyclone Situation Report 1*; 27 November: 2; 28 November: 3; 12 December Department of State, Washington DC.

Ruberu, T S (1976): *Sociological Implications of Malaria and Malaria Control Programme in Sri Lanka* Office of the Superintendent, Anti-Malaria Campaign. Colombo.

Scott, W; Argalias, H and McGranahan, D V (1973): The Measurement of Real Progress at the Local Level; Examples from the Literature and Pilot Study *Report No 73.3* UN Research Institute for Social Development (UNRISD). Geneva.

Sri Lanka Post Office (1978): *Telephone Directory 1978* Part 1: Provincial Exchanges (revised up to 31 December 1977) Postmaster General and Director of Telecommunications. Colombo.

United Nations (1977): *Statistical Year Book 1977* (1976 provisional figures) United Nations. New York.

UNDRO (1979a): *Disaster Prevention and Mitigation: A compendium of current knowledge* Vol 7 Economic Aspects United Nations. New York.

—— (1979b): *Report of the UN Disaster Relief Coordinator on the Cyclone in Sri Lanka 23/24 November 1978 Case Report No 006* Geneva. July.

Visvalingam, T; Black, R H and Bruce-Chwatt, L H (1972): *Report on the Assessment of the Malaria Eradication Programme in Ceylon* Government of Ceylon/WHO WHO. New Delhi. March/April.

World Bank (1976): *World Tables 1976* Social Indicators: Table 6, Holding and Consumption. Johns Hopkins University Press. Baltimore and London.

CASE-STUDY V

Change, and Vulnerability to Natural Hazard: Chiswell, Dorset*

DISASTROUS MANIFESTATIONS of hazard are usually not unique events. In analysing the causes and effects of these occurrences there are problems for analysts, academics and policy-makers in understanding the long-term perspective as the context for recent events and future policies. Understanding will necessarily be made initially more complex by the variety of standpoints of different interest groups in the affected community, and of the community at large.

Physical permanence of a community cannot be assumed in a changing environment. Vulnerability to the sea has increased during the 1000 years of Chiswell's existence, and is continuing to do so. Understanding of this changing state by various groups in society, and their administrators, is the key to the selection and effectiveness of interacting social and technological measures, whether undertaken specifically against hazard or not.

The extent to which technology can be effectively mobilized and implemented to ensure prolonged community permanence may be assessed only by detailed analysis of environmental phenomena on the one hand, and by comparison with social adjustments on the other. Social adjustments cannot be compared until options for them are made available by the authorities elected for their administration.

The condition of vulnerability is not static. Analysis and assessment of short- and longer-term issues is at once a multi-disciplinary process calling for a fusion of physical and earth sciences, social sciences, and political and administrative processes. That these sciences and processes are themselves evolving, and are not static, is as true as for vulnerability itself. That all are in short- and longer-term processes of change must be understood if each is to be integrated with the other for maximum comprehensive and effective response to natural hazards.

Introduction

Chiswell is a community of 134 people. It is situated at the foot of the north-western slope of the so-called Isle of Portland, off Weymouth, in Dorset; adjacent to, but below, the much larger village of Fortuneswell (Figure 1). Portland is an island but for the shingle bar, or tombolo, and causeway which link it to the mainland. The shingle bar, Chesil Beach, extends for 10 miles along the coast of Lyme Bay from a point by Abbotsbury in the north-west to Chiswell itself to the south-east. For eight miles of its total length Chesil Beach encloses a lagoon, called the Fleet, between it and the mainland. For the remaining two miles at the south-east end, the sea is on both sides, Lyme Bay on the one and Portland Harbour on the other. It is where the shingle bar joins the Isle of Portland and

* Formerly published in *The Environmentalist* 3 (1983) pages 277–87

forms a brief trough between the north-west slope of the Isle and the crest of the shingle that the Chiswell community is located.

Flooding by sea water seepage through the shingle, and by seawaves overtopping the shingle bank is frequent in Chiswell. In December 1978 and in

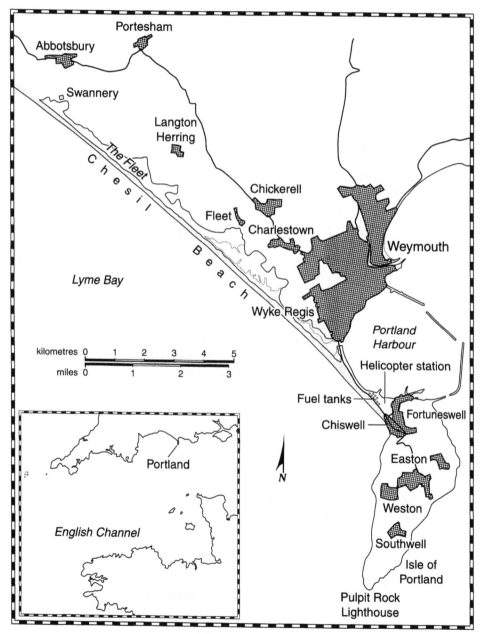

Figure 1

February 1979, for example, waves overtopped the bank with such force that several buildings were damaged, and in February the causeway road serving the whole of Portland was breached.

Geomorphic change

Ridge Height
Chesil Beach itself is between 150 and 200 yards wide, but is narrower both adjacent to the cliffs in the north-west and at the extreme south-eastern end. The ridge of the beach progressively increases in height from the north-west to the south-east, the maximum of 45 feet above mean sea level being found adjacent to Chiswell (Carr, 1978). At the north-western end, the shingle bank had increased in height by about six feet between 1853 and 1969. Over the central area, from Langton Herring to Wyke Regis (see Figure 1) the bank has increased in height by about five-and-a-half feet.

At the south-eastern end, adjacent to Chiswell, there has been a reduction in the height of the shingle of as much as eight feet during the 116 year period (Carr and Blackley, 1974; Carr and Gleason, 1972). Storm waves may periodically destroy the crest, but over a long period there is an average 'equilibrium profile' (Wolman and Miller, 1960) by which the beach may be characterized. This profile is maintained by frequent events of small magnitude, encouraging reversion to the equilibrium profile after change caused by a sudden extreme event.

The crucial question is how the equilibrium profile is to be identified at a particular point in time, or how reliable measurements are to be recorded. Vulnerability of Chiswell to storm and flood is increased by short-term changes as well as long-term evolution.

Assessment of vulnerability to overtopping waves requires recognition of the frequent events of small magnitude and prediction of the rarer extreme events. The rare event which may severely reduce the ridge height of Chesil Beach may increase the vulnerability of the Chiswell community to the next frequent event of small magnitude. It may take only small magnitude events to overtop the previously damaged bank. Even though subsequent small magnitude events will help to put it back again, it is of little comfort to the community to know that within several generations everything will be as it was before people died or lost their homes.

After the beach crest was destroyed by the winter storms of 1978/9, the local authority undertook to re-form the crest by bulldozing pebble material back into place, grant-aided by central government. Can the integration of this action into the geomorphic cycle of events be assured, and can the undertaking be more than a temporary palliative in the face of inexorable natural forces?

Movement
There is additional lateral movement of the shingle bank, although landward recession of the bank may have been more rapid in the past than it is at present.

There is also some longshore movement along the line of the shingle bank accompanied by sorting of shingle size. The size of pebbles broadly increases towards the Portland end and material is transported by tidal currents and waves from east to west (Carr, 1969). There is also vertical sorting, more active at Chesil which is more exposed to open sea. Larger stones at this south-eastern end cause porosity to be greatest. Longshore movement of shingle is probably less adjacent to Chiswell, where the approach of waves is directly in line with the approach of the sea in the English Channel (Carr, 1969).

The direct approach of more frequent waves is the most likely cause of some landward movement of the beach at this point. Movement during the 127 years since 1852 has been estimated as between 42 and 66 feet, and may have been greater in earlier times (Carr and Blackley, 1974; Carr and Gleason, 1972).

There was probably a forerunner of the beach 80 000 years ago, and a Chesil Beach analogous to the present formation from 6000 years ago (Carr, 1978). One hundred, or even 200 years is a small fraction of the several thousand years since the bank formed offshore and began its inexorable mainland recession.

Recession has increased the vulnerability of Chiswell, and especially the causeway at Portland Harbour. The most significant factor of vulnerability at Chiswell is the reduced height of the beach crest, which is now nearer to Chiswell than it was when village settlement commenced. Movement is continuing, and vulnerability of the community at Chiswell to the sea overtopping the beach, and to seepage, is increasing.

Vulnerability

In the great storm of 1824, the church at Fleet village, three-quarters of a mile inland from the sea-line of Chesil Beach, was destroyed with several houses but no lives were lost. By contrast, in Chiswell – five miles to the south-east, 'upwards of 80 houses' were damaged or destroyed and 26 people died (Portland Flooding Sub-Committee, 1979). Most of Fortuneswell, on the Isle of Portland, is between 100 and 150 feet above sea level; but Chiswell is between 10 and 15 feet above mean sea level, and below the 45 foot ridge of Chesil Beach. When waves overtop the ridge elsewhere in its 10 mile length, water runs into the Fleet and eventually to the sea at Portland Harbour; when they overtop at Chiswell they wash immediately upon and into habitation.

Chiswell's unique location in respect of its direct vulnerability to the sea is matched by advantageous proximity for fishermen. Chiswell is the only community on the Isle of Portland with this close proximity, a practical advantage recognized since Roman times. Chiswell was probably established as the principal source of fish for the Island (Morris, n/d.) at a time when the sea was appreciably further away than it is now.

A watercolour of 1805 details the community very clearly as a group of approximately 47 dwellings (Weymouth Local History Museum). Several boats are shown drawn up on the shingle ridge, and the line of the present main street, parallel to the beach, is clearly identifiable, but there are also numerous rows and single dwellings at right angles to the main street on the seaward side, rising up the shingle slope. From a photograph taken probably about 1830 (Bettey, 1970) it is possible to identify approximately 15 dwellings between the main street and the ridge of the beach, which are now no longer standing.

As a principal community on the island, the population and number of dwellings in Chiswell can be assumed to have expanded throughout the nineteenth century, as did the population of Portland as a whole, due to the prosperity of the quarrying industry for Portland stone and employment on the construction of Portland Harbour. Population peaked at 1901 and then reduced by over one-fifth by 1931 (Bettey, 1970). Since then there has been a gradual increase in the population of Portland which has not been reflected in Chiswell. In 1939 there were 33 commercial premises listed in Chiswell (*Kelly's Directories*, 1939) including six public houses or hotels; in 1979 there were 11, including only three hotels or public houses.

The present population of approximately 134 in Chiswell (Chiswell Residents' Action Group, 1979a) occupies approximately 70 dwellings, but in the great storm of 1824 a total of 'upwards of 80' houses were 'damaged or washed down', and in the storm of 1942 100 houses were reported damaged (*Western Gazette*, 1942). It is not possible to determine what proportion of the whole community these figures of damage represent, but Chiswell was only recently very much larger than it is today. The decline of 21 per cent between 1901 and 1931 has been followed by probably 50 per cent decrease since 1942.

There are two other factors to consider before assessing the role of natural hazard in this decline. Close proximity of Chiswell to major naval, and minor civil, south coast ports created considerable vulnerability to damage by enemy air action between 1939 and 1945, and it is reasonable to assume that some war damage occurred in Chiswell. Post-war planning for Chiswell aimed at 'seaside' redevelopment; rebuilding of derelict or extensively war-damaged properties was not permitted (Dorset County Council, 1964). Most of these and other existing properties were considered to be not in character with their seaside setting. The area was to be tidied up, the reinstatement of properties would conflict with proposals of the local planning authority for the clearance of the area between the main street and Chesil Beach.

There had been storms and floods in 1945, 1949 and 1954 (Morris, 1979) but natural hazards were ignored in post-war planning, which focused exclusively on the architectural and visual aspects of development.* The utilitarian quality of much existing construction was considered undesirable, even though it may have resulted from a once indigenous appreciation of sea hazards. Chiswell was to be replanned in the same 'seaside' tradition as Georgian and Edwardian Weymouth.

In the immediate post-war period, damaged buildings were treated as if they were all a result of enemy action. That they included buildings damaged by the sea is certain, just as there are buildings in Chiswell now damaged by the sea in

---

* See Chapter 6: *Institutions and Policies* for historical perspective on war and reconstruction and the policies that ensued; at Chiswell, these are in microcosm.

1978 and 1979, and others derelict and said to have been damaged by earlier storms. It is impossible now to assess the effect of storms on the 'dereliction' and 'blight' that the planners of the 1960s were so concerned about, but storm contributed to the planners' decisions, made between 1945 and 1961, to demolish or close a total of 36 dwellings considered to be sub-standard or unhealthy. There is documentary evidence of the decision (Portland Borough Council, 1962), though not all demolition plans may have been implemented.

'Tidying-up' and refusals for re-development 'on the grounds that such proposals represent a piecemeal approach likely to be detrimental to the possible future development of the area as a whole' had a totally negative result. Areas designated as public open space still contain the foundations and some walls of dwellings demolished by man or sea, and no 'development of the area as a whole' ever took place. Four new houses were built in 1970. In addition to the arrest of spontaneous change, the removal of seaward derelict buildings has increased exposure to the sea for the remaining buildings (Chiswell Residents' Action Group, 1979a). The few remaining inhabited buildings of the rows that extended at right angles to the main street, up the beach, are protected by the long-standing Cove Inn, massive and erect on top of the sea wall and shingle ridge.

Damage to Chiswell by seawater is caused by seepage through the shingle bank as well as by overtopping. In 1824, seawater in the Fleet rose to a depth of 22 feet 8 inches at the Swannery at Abbotsbury (Arkell, 1965). A 100-ton sloop was carried over the shingle crest and subsequently relaunched into Portland Harbour. The higher the sea rises up the shingle bank, the greater the pressure of water to cause seepage through the bank, the lesser is the volume of shingle to obstruct seepage, and the larger components of shingle increase the rate of seepage. Finally, the sea overtops the crest aided by storm waves carried on top of a flood tide. When the sea is high, water seeps up out of the shingle bank on the landward side and runs through and between the buildings and dwellings of Chiswell. Waves hurl stones and pebbles, causing their own impact damage to roofs and windows. Flooding is often exacerbated by heavy rain, which has itself occasionally caused some collection of flood water.

There have been 21 storms and/or floods since the great storm of 1824 (Morris, 1979). In 1942 seepage through the bank commenced an hour before waves began to sweep over the beach. Slight flooding in Victoria Square, the lowest point, then began to become serious, and that area was eventually flooded to a depth of six feet. Waves overtopped the shingle crest, and over 100 houses in Chiswell were flooded and the road and railway (now disused) on the causeway were breached. '. . . tin baths were swept from houses. . .' and the suffering of flood victims was compared to victims of enemy air attacks, being just as deserving of help, but not eligible for assistance from the Lord Mayor of London's Air Raid Distress Fund (*Dorset Daily Echo*, 1942). The storm occurred on 13 December, the reports of damage being delayed for nine days due to wartime restrictions on news reporting.

On 12 December 1978, Victoria Square was flooded to a depth of four feet and high seas breached the causeway for five days. Winds were recorded as Gale Force 9 – with gusts of up to 70 mph, and a section of the ridge of Chesil Beach was demolished. Police issued flood warnings in the early hours and residents adopted 'a now familiar routine of taking emergency action to beat the floods' (*Dorset Evening Echo*, 1978: December 12, 16).

On 13 February 1979, before complete recovery from the December storms had been possible, the sea overtopped Chesil Beach without warning at 6.30 in the morning. Whereas in the December storm, there had been certain points along the ridge where overtopping occurred, on this occasion there was a continuous sea which overtopped a very long stretch of the beach at a height of between 15 and 20 feet. 'This had resulted in instant flooding and there was no action which could have been taken to prevent it' (Weymouth and Portland Borough Council, 1979b). Victoria Square was flooded to a depth of four feet, parked cars were piled on top of each other, electricity and gas mains in the causeway were broken, stones and masonry were swept through breaches in buildings, and 24 people were evacuated from their homes (*Dorset Evening Echo*, 1979: February 15, 15).

The storm which caused the events of December 1978 was local and its results in Chiswell were direct, but the February 1979 sea-surge was a result of a storm or storms in the Atlantic which, with coincident meteorological and hydrological conditions, sent the sea-surge up the English Channel to be trapped by the promontory of Portland. There was no storm at Chiswell when the surge struck, hence the surprise and lack of natural warning when it occurred. Storms such as that in December 1978 are regular occurrences at Portland, with an estimated return period on spring tide of five years, with minor floods as frequent as twice yearly (Dobbie, 1979). However, neither was the sea-surge unique, the previous similar event having been in 1904 and the return period having been calculated as 50–70 years (Dobbie, 1979).

Technological change

In its three-and-a-half square miles, and in addition to Chiswell, the Isle of Portland contains four principal communities, two ancient castles, quarries and stone works for Portland stone, coastguard stations and Pulpit Rock lighthouse, a prison and a Borstal, a hospital, a naval helicopter station, an underwater weapons establishment, and dockyard installations and fuel depot for Portland Harbour. Most of these communities and establishments are elevated and, while exposed to wind, are protected from the sea. Chiswell, almost at sea level, is a very small community incidental to the island's other activities and uses, but its main street is the one-way main road off the island, which passes through Victoria Square, and very near to the naval helicopter station.

The helicopter station was built by the Admiralty in 1962–3 over what had been Portland Mere, which had served the same drainage function as the Fleet still does to the larger section of the beach north-west of the causeway. Earlier maps marked the Mere as 'liable to floods' (Ordnance Survey, 1930) caused by the drainage of seawater from Chiswell. The area was little used, but the construction of the helicopter station blocked the natural escape of excess water.

Although there are two culverts, they become easily blocked with rubbish and debris (Chiswell Residents' Action Group, 1979a).

There is a two-hour difference between high tides to the west and to the east of the causeway, and the Mere served as a ponding area for excess water awaiting the ebb of the tide in Weymouth Bay. Even fully operative culverts require the low tide, and ponding now takes place in and adjacent to Victoria Square. Such was the volume of water trapped in this way in February 1979 that the perimeter five-foot high stone wall of the station was demolished (Chiswell Residents' Action Group, 1979a).

Similarly, the conversion of Victoria Square into a roadway intersection has contributed to successive increases in the general road level as a result of road-works and resurfacing. While this may have had the effect of reducing the depth of flooding in Victoria Square, it further impeded the run-off of excess water from Chiswell. There are houses adjacent to Victoria Square which formerly had up to five steps from their elevated ground floor level to the street, a sensible precaution unheeded by the highway engineers (Chiswell Residents' Action Group, 1979a).

Chiswell's primary long-term vulnerability is from the sea. Secondary vulnerability from deeper and longer-lying floods has been brought about by technological changes over shorter time periods. These changes have been brought about by, or on behalf of, the numerous and significant authorities and institutions that have adopted Portland as a base, and the thousands of people who live or work there.

A sea-wall and esplanade was constructed in 1962 for 1600 feet from where Chesil Beach runs into the north-east face of the island at Chesil Cove, to a point in line with half-way along Chiswell. Construction was then considered possible only with foundations on clay; construction on deep shifting shingle was more difficult, which accounted for the apparently arbitrary end of sea-wall protection. Protection afforded to Chiswell by the sea wall is evident in so far as its limited height and extent allow. Since the first publication of this study, the sea wall has been extended.

The construction of the sea wall reflects acceptance of responsibility by the authorities concerned for the safety of Chiswell, but the wall is overtopped from time to time, its design height being lower than the adjacent natural beach crest. It has been observed that the construction of the sea wall may be a cause of the natural reductions in height of the adjacent beach crest (Carr and Gleason, 1972).

Social change

The authorities that have a direct responsibility for storm and flood hazard and its consequences are the Wessex Water Authority, which has responsibility for coastal sea defences, and the Weymouth and Portland Borough Council (amalgamated in 1974), which has responsibility for evacuation, rehousing, relief and road maintenance. Both these authorities are able to apply to ministries within central government for financial assistance, subject to the respective minister's approval within certain proportional maxima. Both authorities have a responsibility for the removal of excess surface water.

Non-governmental organizations involved with raising and distributing relief funds at Chiswell have been the local Rotary and Lions Clubs, and the Round Table. Most recently, as a direct result of the December 1978 and February 1979 floods and storms, the Chiswell Residents' Action Group was formed (Davey, 1980).

Consultant engineers were appointed by the Wessex Water Authority after the floods of 1979 to assess all available data relevant to flooding at Chiswell, to advise on probable return periods of the storms and floods, on the necessity of further studies, and to suggest options with budget costs, to 'safeguard' Chiswell from flooding.

All the four measures considered in that report are of engineering construction to prevent flooding by seepage, over-topping, and to reduce the energy of sea waves. The important need for a warning system is emphasized, but in this regard 'many problems remain to be solved'. The report concluded with a recommendation for further studies and the preparation of a 'full feasibility report', estimated then to cost £150 000 (Dobbie, 1979).

The Council's activities in the meantime have focused on the drainage of flood waters, the establishment of an emergency control centre, the use of earth-moving equipment to replace shingle from the rear of the beach to re-form and maintain the ridge height, and to liaise with owners where property has become unusable. A special sub-committee of the Council's Policy and Resources Committee was appointed to consider the problems of flooding, necessary remedial action, and the future of the Chiswell area. Initial priority of concern and the emphasis of measures taken has been on the construction of physical measures to resist the forces of sea and storm. The need for warnings, which received secondary mention in the engineers' report, has been realized and approached as a matter of co-ordination between local authorities and water authorities with responsibility for Chiswell, and those responsible for other areas further westwards (Weymouth and Portland Borough Council, 1979a).

It was not until 12 April 1979 that consideration was given to Chiswell residents' housing difficulties resulting from the floods (Weymouth and Portland Borough Council, 1979a). As a result, it was agreed that 'positive moves' should be taken with regard to these properties as a first step towards regenerating the area, and to ensure that the community is revitalized. The Department of the Environment (central government) was to be approached by the Council for possible financial assistance.

Serious hardship was being experienced by some Chiswell residents unable to live in flood- and storm-damaged properties upon which they were committed to mortgage repayments. Relief on these payments had in some instances been given for three months, an insignificant period of time in relation to the scale of the damaging events experienced.

While these conditions may not always be directly connected with local authority administration, there is clearly no policy for their general consideration, nor their co-ordination. Neither is there, it would seem, any policy for engaging in the preparation of schemes for physical protection and prevention of flooding. Decisions for taking such measures are based on an acceptance of moral responsibility rather than as part of a specific policy declared as a result of comprehensive problem analysis. Consideration of social measures and the preparation of flood and storm warnings, and the consideration of means for property purchase and compensation, have been secondary, and in the latter case are a result of instigation by residents.

The formation of the Chiswell Residents' Action Group (CRAG) itself was partly an expression of frustration and concern due to the absence of a stated policy on social measures. In fact, were a policy for property purchase and compensation to be introduced, a larger proportion of Chiswell residents would have left the area. Local estimates were between 20 and 50 per cent in addition to some who had already left. In May 1979 there were 30 recently vacated domestic and commercial properties in Chiswell, and four were for sale. Social measures cannot be left to the relief funds initiated and managed by voluntary organizations. The impressive £8000 total of the fund for the two recent floods made possible the allocation of only £140 each to 50 households, with £1000 remaining in the fund for its recommencement after the next flood.

CRAG co-ordinated and mobilized local residents' opinion, mobilized action for which individuals may have been, or felt themselves to have been, ineffective, produced an 'Analytical Report' concerning the flooding problems, explored possible avenues for compensation or relief, such as discussing possibilities of reduction in property rates, requested and achieved representative co-option as non-voting members of the meetings of Weymouth and Portland Borough Council, and formally applied to the Disaster Fund of the European Economic Commission for financial aid for the alleviation of flooding at Chiswell (Chiswell Residents' Action Group, 1979 a,b,c; Dorset County Council, 1964). Although the authorities would claim that they would have similarly attended to the problems caused by flood at Chiswell whether or not there had been a residents' action group, the Group's activities were taken seriously; the authorities acting in some cases only after approaches had been made to them by members of the Group.

The Wessex Water Authority pointed out that they are not *obliged* to protect communities from flooding. The Weymouth and Portland Borough Council was 'firmly of the view that Chiswell must be preserved, and indeed enhanced' (Portland Flooding Sub-Committee, 1979). CRAG stated their objectives to include lobbying for 'a scheme or schemes that will end for all time the danger to Chiswell from flooding' and 'to ensure that the environment of Chiswell reflects the expectations of the people who reside there' (Chiswell Residents' Action Group, 1979a). All these statements were made before comprehensive analysis of vulnerability to flood had been completed.

The Residents' Action Group balances the absence of social measures by the authorities for flood alleviation. However, neither the authorities nor the Action Group have a policy formulated on analysis; both have discharged what they saw as their respective duties on the basis of moral concern after flooding had occurred. It is likely that discussion and negotiation between the two bodies may actually have impeded analytical processes and official policy formulation. The role of CRAG could only be that of a grassroots pressure group, whereas that of the authorities is long-term. Had there been a policy by the authorities to include social measures at the outset, the formation of the Action Group may not have been necessary.

Policy options

There are no figures available for the cost to the local authority of flood emergency services and repairs to roadways and property. The total cost of damage in the storm of 1942 was put at 'several thousand pounds' (*Dorset Daily Echo*, 1942) and in 1972, after storms in February of that year, the total value of 45 damaged dwellings was put at £330 000 (*Western Gazette*, 1972). Total damage from the February 1979 storm has been estimated at £250 000 (*The Observer*, 1979). These figures are estimates of total damage to both property in local government ownership and that in private ownership.

To these costs must be added the value of voluntary relief funds which have been established, either nationally, as in 1824 (*Morning Chronicle*, 1824; Morris, n/d), or locally, as more recently by the Rotary and Lions Clubs, and the Round Table. The joint fund raised by these three bodies in 1978 and 1979 totalled £8228 (*Dorset Evening Echo*, 1979: February 15, 15).

The cost of the sea wall in 1962 was £180 000 and investment into the improvement of property since the Housing Improvement Act of 1962 could have been £1000 for say half the dwellings in Chiswell, and have totalled around £35 000. Were it necessary for central government grant-aid sources to apply rules of cost effectiveness for proposals submitted to them for sea defences for Chiswell alone, it is difficult to see how they could justify their potential 65 per cent proportion of the total cost of six million pounds.

Were sea defence not to materialize, the inhabitants of Chiswell would have to reconsider their collective and individual alternatives; either to accept continued and increased risk, or to move away in the hope of compensation for loss of property or property value. It is this last option which should have been safeguarded in the first instance as a matter of policy. Only when the size of the remaining community is known can any cost effectiveness be assessed for civil engineering preventive measures.

The fishermen who formerly lived in Chiswell accepted in the 1960s recently completed housing rentable from the local authority. Had vacated properties not been reoccupied by newcomers, recent problems would not have occurred to such an extent. Since the Housing Improvement Act of 1962, assistance has been available for owner–occupiers to improve their properties. Where central and local authorities are prepared to be involved in domestic improvements, for the purpose of improving national housing stock; they must surely be prepared to be involved at the domestic level, where financial inducement for domestic improvements may have encouraged continued occupation of property in a hazardous location.

In any case, involvement of the authorities on the one hand may encourage a reliance upon the authorities on the other. The authorities must, therefore, through planning processes, be as aware of natural hazard situations as private individuals might be expected to be. Where there is a potential partnership on the one hand between the state and the individual, a partnership cannot be avoided on the other. Both are equally in the interest of the state.

There appears to be every reason for developing a scheme to secure the same option of domestic mobility in a situation of hazard, as is enjoyed by members of communities subject only to normal supply and demand of dwellings. The formulation of this kind of policy for social measures should have priority over the application of physical constructed preventive measures. Legislation for such measures would counterbalance the long history of legislation that exists for sea defences and flood prevention (e.g. Coastal Protection Act, 1949; Land Drainage Act, 1976).

However, the fusion of official purposes that such a state of affairs implies will not be achieved through sectoral separation in administration. The planners of the 1960s demonstrated significant social measures which would have more positive results where integrated into a comprehensive policy. Not only was Chiswell diminished, but vulnerability was increased for the community that remained.

In these times of energetic concern for environmental conservation, it could now be proposed that Chiswell be protected as a designated Conservation Area. Chiswell was apparently a visually attractive place at the turn of the century and after (Morris, n/d), but has since suffered from enemy action and post-war planners – as well as storms.

Preservation will not serve to hold back the encroaching shingle, however, as has been suggested (Chiswell Residents' Action Group, 1979a). The shingle has been shown to be advancing inexorably landwards and the vulnerability of Chiswell has increased for that, and other reasons. People and community are a part of the environment, not separate from it. The sea and shingle, which once created opportunity, taken by people, for advantageous proximity to the sea, now choose in time to take that opportunity away. Practical advantages, once predominant, are being slowly supplanted by the disadvantages of hazard, to which conservation would in this case prolong exposure.

Preventive measures against hazard must therefore take comprehensive account of the relationship of man and his environment, and must be ecological adjustments in the activities of vulnerable people and their elected administrations, rather than only separate technological resistance to the forces of hazard (Lewis, 1979).

Ecology in this case is the relationship of society, via its adopted administrative processes, with its social and political environment as its means of effective and comprehensive relationship with its physical environment. That some of society's options with regard to the physical environment are administered and controlled by its administrators must be understood by those administrators. Vulnerability is compounded of physical and social conditions, and preventive measures must be compounded of physical and social measures.

The preparation of warnings and their dissemination are of prime importance in measures for preparedness. Advice on hazards to be expected, on what to do, on how to secure property against flood, on what evacuation procedures will be available, how and where to make contact with the authorities, and what measures various authorities will be undertaking in the event, are all examples of preparedness planning (Lewis, 1977b). It is obvious that preparedness planning is multi-disciplinary and multi-sectoral, and calls for the closest integration of measures to be taken by authorities and the public at domestic level.

One overriding additional factor has implications for policy formulation in respect of natural hazard at Chiswell. Twelve-and-a-half-thousand people live on Portland, some commute to the mainland and probably several thousand more commute from the mainland to the institutional, scientific, military and commercial establishments on the island. They are all served by the causeway road and by the one-way road approach and exit system to the island which includes the main street of Chiswell. The causeway road and the electricity, gas, water and telephone utilities and services under it, are all afforded protection from the sea by Chesil Beach; as are the naval fuel tanks, and the naval helicopter station and the western side of Portland Harbour.

Vulnerability to the sea is increasing for these institutions, as it is for Chiswell. This may make all the difference to considerations of cost effectiveness for sea defences, or to the possibility of obtaining financial assistance for them. In this case, however, the danger to Chiswell may be from increased secondary hazard of floodwaters. There is also the added danger of preventive measures designed for longer return periods permitting and encouraging development of Chiswell, which may bring about larger disaster on the rarer occasion of eventual overtopping.

Were it not for special consideration for the Island as a whole, policies by the local authority for the preservation and enhancement of Chiswell would have brought greater demands for protection by the water authority. As central government is involved in the improvement of housing stock and sea defences, one hand pays for the protection of what the other hand creates.

Similarly, the extraction of pebbles was continuing in 1979 some 25 miles along the Dorset coast at West Bay. Removal of aggregate was licensed by the West Dorset District Council, which received a royalty per ton. At the same time expenditure was being incurred by nearby authorities to combat coastal erosion and flooding (Carr, 1979).

Postscript

In October 1980 judgement was delivered in the Queen's Bench Division against the local authority responsible for road and bridge construction which had caused flooding for a single house owner near Dorking. The house, built in the 17th century, had not been flooded before 1900 when the bridge had been built and road levels changed. Culverts installed at that time were inadequate and occasionally became blocked. Judgement found that flooding was substantially caused by an inadequate culvert and the 'damming effect' of the bridge and its approaches. Flooding was both a public and a private nuisance. Damages of £3300 were awarded to the plaintiff and works were to be undertaken to 'discontinue the nuisance' (Law Report, 1982).

Acknowledgements

This study is an edited version of *Vulnerability to a natural hazard: Geomorphic, technological and social change at Chiswell, Dorset* published as Natural Hazards Research Working Paper No 37 in December 1979 by the University of Colorado. The postscript updates it to 1982.

The author wishes to acknowledge, with thanks, the assistance of the following in the preparation of the earlier paper: Mr K G Anderson of the Wessex Water Authority; Mr Stuart Morris of Weston, Portland; and Mr Rhys Davey, Chairman, Chiswell Residents' Action Group. Special thanks are due to Dr A P Carr of The Institute of Oceanographic Sciences for his comments on a draft version of the paper.

References

Arkell, W J (1965): The effects of storms on Chesil Beach in November 1954 *Proceedings of the Dorset Natural History and Archeological Society* 76 pp 141–145.

Bettey, J H (1970): *The Island and Royal Manor of Portland: Some aspects of its history with particular reference to the period 1750–1851* The Court Leet of the Island and Royal Manor of Portland in association with University of Bristol, Department of Extra-Mural Studies.

Carr, A P (1969): Size grading along a pebble beach: Chesil Beach, England *Journal of Sedimentary Petrology* 39 (1) pp 297–311.

Carr, A P (1978): The Long Chesil Shingle *Geographical Magazine* 50 (10) pp 677–680.

Carr, A P and Blackley, M W L (1974): Ideas on the origin and development of Chesil Beach, Dorset *Proceedings of the Dorset Natural History and Archeological Society* 95 pp 9–17.

Carr, A and Gleason, R (1972): Chesil Beach, Dorset and the cartographic evidence of Sir John Coode *Proceedings of the Dorset Natural History and Archeological Society* 93 pp 125–131.

Carr, A P (1979): Correspondence *Institute of Oceanographic Sciences* 10 September.

Chiswell Residents' Action Group (1979a): *An Analytical Report Concerning the Flooding problems of Chiswell, Portland* Mimeo. 10 pp. March.

Chiswell Residents' Action Group (1979b): Letter to the President of the European Commission 11 April 1979.

Chiswell Residents' Action Group (1979c): *CRAG Bulletin Nos One and Two* March 1979 and 23 April 1979.

Davey, R I (1980): Attempts to prevent flooding at Chiswell, Portland *Disasters* 4 (4) pp 380–382.

Dobbie, C H and Partners (1979): *Preliminary Report on Flooding at Portland* Report to the Wessex Water Authority, Southampton. March 1979 (revised).

Dorset County Council (1964): *Report on Chiswell Area, Portland, (with suggestions for improvement and policy with regard to development and redevelopment)* South Area Planning Office.

Dorset Daily Echo (1942): 22 December.

Dorset Evening Echo 22 June 1972, 4 April 1977, 12 December 1978, 13 December 1978, 14 December 1978, 15 December 1978, 16 December 1978, 13 February 1979, 14 February 1979, 15 February 1979, 2 March 1979, 19 March 1979, 1 April 1979. Weymouth, Dorset.

Kelly's Directories Ltd (1939): *Kelly's Directory of Dorsetshire 1939* London.

Law Report Queens Bench Division: Potter and Others v. Mote Valley District Council and Another, before Mr Justice French *The Times* October 22, 1982. London.

Lewis, J (1977b): *A Primer of Precautionary Planning for Natural Disaster* Disaster Research Unit, University of Bradford. February.

Morning Chronicle Thursday, 30 December 1824. London.

Morris, S: (n/d) *Files on the History of Chiswell.*

Morris, S (1979): *Resume on the History of Flooding at Chiswell since 1824* Mimeo, 3pp.

Observer 19 March 1979. London.

Ordnance Survey (1930): *Isle of Portland* 6"–1 mile: Dorset Sheet LVIII SE. Southampton.

Portland Borough Council (1962): Derelict Buildings, Demolition Orders *Minute No. 799.*

Portland Flooding Sub-committee (1979): *Minutes of Meeting 6 April* (including Report of the Borough Engineer prepared 1939, by Morris, S.). Weymouth and Portland Borough Council, London. Weymouth.

Western Gazette: 30 June 1972, 9 March 1973, 13 February 1976, 23 February 1979, Undated 1942. Weymouth, Dorset.

Weymouth and Portland Borough Council (1979a): Official Notice and Reports *Minutes of Meeting 15 March* Weymouth.

Weymouth and Portland Borough Council (1979b): Official Notice and Reports for 15 March *Policy and Resources Committee* pp 284–285. Weymouth.

Weymouth Local History Museum: *The Common of Chesilton* John Williams Upham, c1805. Watercolour (Ret No PI 17) ('John Penn' Series No. XVII).

Wolman, M G; Miller, J P (1960): Magnitude and frequency of forces in geomorphic processes *Journal of Geology* 68 (1) pp 54–74.

PART THREE
A PATTERN FOR DEVELOPMENT

The physical environment presents itself as it is; no more, no less. It is society that ought to learn and constantly reduce its limitations with respect to the physical environment. Famine is a consequence of the failure to learn from the constant interactions between a society and its physical environment. In these constant interactions the burden of adjustment is on society, not on the physical environment. Famine is a human responsibility (Wolde Miriam, 1986).

As are earthquakes, tropical cyclones, volcanic eruptions, fires and floods.

6

Development and disasters

A brief history

INFORMATION THAT REACHES us about natural disasters, focuses on those disasters of large magnitude. This fact has led in the past to the assumption that disasters, though catastrophic, were comparatively rare. In their rarity, therefore, we could also assume that their occurrence was outside of normality.

This has meant that for follow-up activities to be publicly credible, those too would best be associated with those disasters of large magnitude that had reached public and popular awareness. It also meant that relief measures would also be on a large scale; anything else would be meaningless. These assumptions have not only influenced governmental and non-governmental measures for post-disaster assistance, but have also influenced disaster research.

The assumption that disasters were abnormal led to their disassociation from everything else that was normal. Day-to-day normality could continue in the knowledge that disasters could occur, but so could the assumption that, as there was nothing that could be done about them, neither were changes to normal day-to-day affairs necessary.

The few exceptions proved the rule. Systems for warnings of storms were arranged, and plans were made for operation in the event, but these were in themselves separate and for rare and occasional use. When not in use, they were put away. Their operation and management depended upon what resources were to hand; it did not make new demands for unusual or special equipment, and had little effect on normal day-to-day affairs.

Perception of disaster as being dissociated from everything else until it happened has had serious consequences. It has meant that activities of settlement, construction, production and commerce could proceed undeterred by the possibility of natural disasters, but that when disasters did occur these activities would have a significant bearing upon their consequences. Moreover, when this happened it was unfortunate, but no one was at fault.

Later on, research indicated that adjustments were necessary so that the impact of natural hazards might be lessened. Adjustments were identified *after* the event, but if implemented would be in readiness for another. Adjustments were indicated in a socio-economic context where options and resources for their implementation were available – although availability did not always mean that adjustments were implemented.

Even later, it came to be understood that what was happening in settlement, construction, production and commerce, dissociated as they were from any awareness that there might have been, was having a far greater effect on the consequences of natural hazards than any adjustments could possibly have. Though adjustments were identified, their implementation depended to a large degree upon either legislation or self-preservation. Meanwhile, activities could proceed in other places where these constraints were less relevant or non-existent.

Development and its administration proceeded in those other places because they were less fortunate, less skilled and less rich. At the time, they were less well informed and accepted development because of external pressure and internal aspirations. Development that might have been 'impeded' at home by concerns inclusive of natural hazards awareness, proceeded in other locations regardless, uninterrupted by cultures and concerns that were different, less articulate in a form that could be understood – and therefore less powerful.

In the meantime, the 'disaster relief' lobby has itself become a large and growing industry; humanitarian assistance has increased many times over, in only a short period, to its present four billion dollars (Madrid, 1995). Humanitarian assistance is growing while development assistance is declining (*The Economist*, 1996; also quoted in Dommen, 1996). Nevertheless, 'humanitarian assistance' struggles to cope with disasters that are reportedly larger, more numerous and more complex but for which, in spite of all its resources, it can perhaps be a palliative but never a long-term solution.

'Complex disasters' are a symbol of our time. Without seeking to deny the crucial need for assistance, as deprivation nurtures violence, it does seem appropriate to observe that deprivation, on the one hand, has been created by activities which, on the other, have proceeded for too long regardless of socio-environmental consequences. 'Simple disasters' have been superseded by ' complex disasters' – but the 'simple' ones have not gone away.

Neither are 'simple disasters' as simple as they were, especially as here, also, the consequences of unconcerned activities in the past have now become the cause of our concerns for the future, and here, also, humanitarian assistance – though necessary – cannot be left to become a substitute for socio-economic and political preventive action (Madrid, 1995).

Linkage between development and disasters has to be forged, if only because the effects of either are not to impede the other. Which way round the effects are most felt, is a matter of operational viewpoint or political bias! The 'disaster continuum' approach positively attempts to align post-disaster assistance with development, recognizing the intervening stages of recovery, rehabilitation and reconstruction. The one, it was intended, should lead to the other – in that sequence.

Disaster, reconstruction and development, are conventionally thought of, perceived and represented linearly. In reality however, they are simultaneous, each 'stage' overlapping with others in the same or neighbouring places, and in responses to the same or different disasters and by the same or different authorities and organizations. In any case, development will have been taking place, appropriately or inappropriately, planned or unplanned, successfully or not, for a long time.

Which way round should it be; 'disasters and development' or 'development and disasters'? Which comes first and which has the greater influence upon the other? The economic impact of disasters, especially on island states, has been a matter for concern and analysis (e.g. Lewis, 1991b) and the impact upon development of disasters is frequently deplored (e.g. ECLAC, 1987; 1988) – but what of the impact of development upon disasters?

A cyclical concept

For a single disaster event, disaster, reconstruction and development are in fact cyclical. Around several events the cycles overlap. Though a disaster may commence a process that leads to development, development has invariably preceded disaster and had a bearing on the extent and implications of the disaster that ensued – for better or for worse.

It has been the practice to refer to 'the disaster cycle', in which disaster, seen usually as the trigger for everything else, occurred in a cyclical sequence of action back to disaster again, acknowledging that disasters could recur; thus: disaster; relief; rehabilitation; reconstruction; preparedness; disaster:

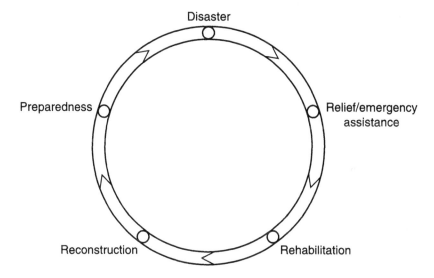

This 'disaster cycle' still appears in descriptions of programmes and projects undertaken on account of natural disaster occurrence.

What the self-centric 'disaster cycle' did not acknowledge was that there were other sectors of activity continuing outside of the cycle. Not everything that happened, or that was undertaken, subscribed to this interpretation of natural disaster management, though it invariably did subscribe to the contexts for the impacts of natural disasters themselves. Development, or simply 'change' was also taking place – of its own inevitable volition or in a planned and programmed way. Where did 'development' appear in the 'disaster cycle'?

In fact there is not one 'cycle' but two; it was not a 'disaster cycle' but a 'disaster bicycle':

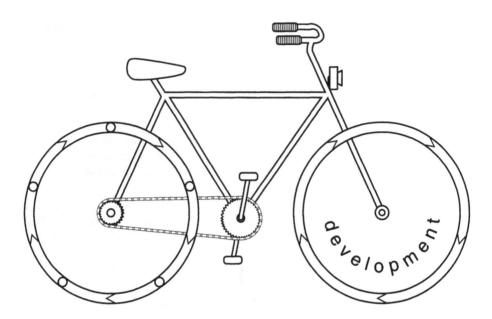

However, the bicycle does the dual system credit that it does not deserve; most bicycles have one rider and one person to do the steering. Whereas, the disaster cycle and the development cycle were not, and still are not, driven by the same authorities.

Of overriding concern is that the activities of this cycle of events are implemented by various sectors and are rarely co-ordinated or objectively interrelated. Activities within the development spectrum are organized and managed separately by their sector. Disaster management has become separated – as another sector – from all of these already separated activities. Of necessity, measures for disaster reduction are widely multi-sectoral and its management is a pervasive part of local experience and context. Disaster management has, however, become separated from other aspects of development, as they are separated from each other. Most crucially, disaster management has become separated from the development of everyday affairs that create vulnerability.

As a result, the activities of one sector may not necessarily subscribe to disaster or vulnerability reduction, which has been made the responsibility of another sector; and by ignorance of processes that subscribe to vulnerability, may actually be making things worse. Vulnerability has frequently been made, or made worse, by 'development' (See Chapter 4).

Questions remain, therefore, to do with the relationship between disasters and development (Cuny, 1983). What kind of development made things worse and, more constructively, what kind of development would have made things better – and will make things better in the future? More precisely – without waiting for disaster to start a process that could lead to development, what kind of development is required in the first place so as to achieve disaster reduction (Lewis, 1996)?

Institutions and policies

The United Nations Relief and Rehabilitation Administration (UNRRA) was formed in 1943, towards the end of World War II. After the end of a war so catastrophic, no other response was conceivable. The urgent and appropriate need for relief, made ridiculous any concept of how relief might become less necessary in the future, or consideration of how development might reduce vulnerability and make survival more likely. Recurrence was inconceivable and issues to do with vulnerability would have been irrelevant in an idealistic world.

UNRRA was followed in 1944 by the International Bank for Reconstruction and Development (The World Bank). The sequence was framed in response to an unrepeatable war that started it off; but was it appropriate to apply the same thinking to naturally recurring disasters (Morse, 1977; Jackson, 1986; Lewis, 1991a)? What was appropriate where recurrence was then regarded as out of the question, was not appropriate where recurrence was in the natural order of things and where there may be the need as well as the opportunity to take account of that recurrence.

The European Bank for Reconstruction and Development commenced in April 1991; the adopted sequence is as customary now as it was fifty years ago: *relief, rehabilitation, reconstruction and development.* (See Case-study V for this international state of affairs to be reflected in microcosm at Chiswell, Dorset, where post-war reconstruction ignored natural hazards).

There continue to be relief agencies and development agencies. There is the United Nations Office for the Coordination of Humanitarian Affairs (OCHA), formerly DHA and the United Nations Disaster Relief Office (UNDRO), and there is the United Nations Development Programme (UNDP). The recently appointed United Nations Secretary-General has proposed a reorganization of the United Nations which maintains a distinction between two groups of aid agencies, one for humanitarian affairs and the other for development. The two sectors are referred together only as a sequence commencing with disaster and ending with development; e.g. 'The Continuum from Relief to Development' (Westgate, 1996).

Simply in terms of size, war and natural disaster are less different than they were at the end of World War II: wars are smaller and natural disasters are said to be larger. In terms of recurrence, both wars and natural disasters might be regarded as being in the same indeterminate but recurrent category. There would now seem to be reason to re-examine the sequence, whether it was assumed or calculated, and on behalf of both natural disasters and human conflict in all their interrelated complexity.

There is certainly no longer any reason or excuse for post-disaster assistance to be separated from development in the same sequential pattern (if linked at all) as it mistakenly was in 1943 – more than half a century ago. There is now even more reason for reconstruction and development to accommodate reductions in both physical and socio-economic vulnerabilities to hazards of all kinds in uncertain futures. Disasters do not belong exclusively to 'disaster relief' and neither does the disaster response sequence have to wait for disaster itself to start it off.

Vulnerability is pervasive; 'vulnerability' has significance beyond disasters discourse. Socio-economic vulnerability to the effects of war and conflict is socio-economic vulnerability to natural disaster, and vice versa. Development to reduce socio-economic vulnerability to one will reduce socio-economic vulnerability to the other. Basic-needs development appropriate to vulnerability reduction and survival serves also to increase the quality of life between disasters. Where conflict is caused or exacerbated by perceived inequalities between populations and regions, equitable basic-needs development may also be the commencement of a process that renders conflict and civil strife less likely.

As it is, and appears to want to be, international disaster response and its divisions remain heavily institutionalized. Departments, offices or units for 'disasters' (DfID), 'humanitarian assistance' (Canada), 'disaster assistance' (USA), 'disaster preparedness' (Jamaica), 'protection civile' (Algeria) for example (UNDRO, 1987), have become their national centres for a conventional focusing of attention on post-disaster effects and the rehabilitation of physical and social conditions affected by specific events.

In answers to a question addressed to European Union member states (Commission of the European Communities, 1996b) on whether a single organizational structure exists, or whether different ministries or departments are responsible for different aspects of humanitarian and development aid, seven replies stated 'different departments', five indicated a single department (and one 'mainly' and one 'decentralized co-operation') with one reply outstanding. Of the 15, 10 stated different budget sources for each type of aid (with three single budgets, and two replies outstanding).

Based upon a *post facto* disaster-specific approach, the predominant system excludes the possibility of a necessarily wider view to take account of crucial political, institutional, social, cultural, economic and physical factors that are the root causes of vulnerability to 'natural disasters'. Institutional separation has taken 'disasters' away from everything else, and has thus implied the absolution for all other sectors of their responsibilities in that respect. It has been encouraged by, and has followed, international inducement and format and has not always been developed on the basis of indigenous needs and conditions.

Consequently, as a result of popular interchange of terms and labels and of common misunderstanding, 'disaster relief' has been assumed to be the totality of necessary action for disasters and not to involve any but the department of that designation. Nothing could be further from reality and there could not be a more dangerous outcome!

Not only does institutional separation reflect shortcomings in the understanding of the crucial relationship between vulnerability, disasters and development – but it denies the opportunity that integration would offer for strategic development for vulnerability reduction. Those preventive measures that are being implemented are largely mono-sectoral, mono-disciplinary, mono-disaster 'type', and largely technological (or 'technocratic'; viz Hewitt, 1983; Winchester, 1986).

As it has become more and more institutionalized, disaster response has become more obviously separated in its management from those sectors responsible for socio-economic development in which ferment the causative factors of vulnerability. Separate sector policies may negate the possibility of holistic strategy to the extent that some sectors may inadvertently be the root cause of disaster consequences that relief sectors of the same governments and organizations are then called upon to attend to, and to pay for (Lewis, 1986).

Institutional and organizational separation is the greatest single impediment to the integration of development and disaster reduction. Separation and rivalry between organizations established to implement various objectives that subscribe collectively but piecemeal to disaster reduction, is inefficient and counterproductive. Neither internationally nor nationally is there one organization or sector that has the prerogative to oversee or to review, but not at all to implement and to encourage, the linkages between development, disaster responses and vulnerability reduction.

As an aspect of the Rio Declaration (1992), multi-sectoral programmes are required for the co-ordination of policy formulation, information sharing, information gathering, analysis and research, promulgation of information to the public and its specialist sectors of construction, agriculture, forestry, food, water and fuel production and supply, communications and transport, training and education. Such programmes cannot be repeated and continued often enough in order recurrently to include representatives at all levels of all sectors in all their diversity of experience, skills, location, context and activity.

Disaster reduction requires the support of, and subscribes to, development. The next step is for development itself to make itself the paramount disaster-reduction activity. Development is not sacrosanct; it has to be moulded to adapt to requirements. The degree by which populations and their activities can absorb, reduce, or change their vulnerability, accommodate risk and achieve survival and recovery, is an expression of prevailing political, social, cultural and economic conditions – which are the objectives of development.

Management for comprehensive development

Quarantelli (1985) has argued for disasters to be regarded as 'social events' (see Introduction and Chapter 1) and points out that if disaster management focuses exclusively on the perceived primary cause or origin of disasters, this will inevitably lead to a misunderstanding of post-disaster problems – which in turn may have important consequences for disaster victims in the long term. Solutions, he suggests, may reside in changes of public policy and in interventions aimed at changing aspects of social structure.

The same kind of observation could be made with regard to pre-disaster initiatives for vulnerability reduction as it has been with regard to the reduction of post-disaster trauma. The observation is made in a perception of sequential stage-by-stage recovery subsequent to a specific event. What happens in reality is that conditions of post-disaster trauma (to continue the example) will become a part of the social context for subsequent and recurrent disasters if interventions, modifications and adjustments to institutional perceptions and divisions are not achieved.

The automatic adoption of existing and conventional management frameworks into which to slot a comparatively recent management function for 'disasters', has had only partial effectiveness and may already have led to an overall negative condition. Existing management functions must themselves be reassessed as extensions of analyses of vulnerability and its contributive processes within its environmental, developmental and institutional contexts.

Theoretical division of concepts by 'outsider' international agencies, was responsible for separate initiatives for 'preparedness', 'relief', and 'prevention'. Separation meant that they could be alternatives, but for successful disaster reduction all these measures have to be comprehensively applied and integrated with each other and within other sectors, as well as with the variety of indigenous normal, as well as extreme, conditions they are intended to serve. For their achievement, they need to be applied locally so as to serve conditions in varying degrees of development evolution. They require indigenous fusion, not exogenous separation, and they require a developmental context in order to serve integral experience of all hazards, not only unique or extraordinary events. A multi-disciplinary, comprehensive, human–ecological and locally integrated approach by indigenous authorities and organizations will be more effective for disaster mitigation than partial, sectoral, mono-disciplinary, policy separation by exogenous agencies.

Processes occurring outside of disaster management strategies have for too long been disregarded by the disaster set. Thus, for example, migration and its causes, squatter settlements and the occupation of hazardous sites, were outside the remit of disaster management, though these processes have created more people vulnerable to floods, landslides, storms and fire and have placed greater demand upon relief and rehabilitation. While it can be argued that there is now, more than ever, the need for these services, it is illogical that demand for them should be increasing without consideration of the causes that make them so necessary. Their increasingly high cost at least, should be an indicator of the need for reassessment. Vulnerability has to be reduced by strategic processes within development; not by the *ad hoc* application of services that commence only after a disaster has occurred, or with the identification of, and attendance to, already 'vulnerable groups'.

The 'continuum from relief to development' has recognized the significance of development itself in the sequence of disaster responses, largely because of the negative impact upon development that disasters exert. Development continues to be regarded as the goal when everything else has been accomplished; relief and rehabilitation are expected to modify so as to integrate with and not to impede or counter development (as numerous titles testify, e.g. *Linking relief to development: Disaster response with foresight*: OFDA, 1997a; *Linking relief and development*: IDS, 1994). There has not, as yet it seems, been much consideration of how development itself could be modified to anticipate and to accommodate disaster and disaster responses; above all, to achieve reductions in vulnerability to disasters and to link development to a reduction of the need for relief (but see *Linking development to relief: The integral role of prevention, mitigation, preparedness, and planning*: OFDA, 1997b).

It is interesting to observe how long it has taken to reach even the beginnings of a recognition of a continuum – there has been half a century of natural disasters since the establishment of the United Nations Development Programme. Research initiatives have not been wanting (e.g. Lewis, 1982a), but it has taken this time for the relief and rehabilitation lobby to achieve sufficient power to meet the development lobby on something like equal terms. Development has suffered not only from disasters, but from conflicts of funding; as demands for humanitarian assistance increase, capacity for development has fallen in a crisis-driven response system. This at a time when it is more than ever self-evident that what is required is the identification and modification of causes, as well as attention to the symptoms (Clarke, 1997).

A significant and comparatively recent observation has been that because the operations of relief, rehabilitation and development may all be occurring simultaneously within any country or place, 'disaster continuum' is an inappropriate term, implying as it does that each phase follows the other in sequence. 'Disaster contiguum' has been expressed as conveying a more accurate recognition of a more normal reality (as a footnote in CEC, 1996b; see also Chapter 5). In the same document, a role for development is identified (in another footnote) 'to prepare for risks or prevent disasters'. The objective of the communication is plainly the achievement of a political rather than a functional balance between three separate phases of operations:

> Better development can reduce the need for emergency relief; better relief can contribute to development; and better rehabilitation can ease the transition between the two.

The three phases of operation were represented (no doubt) by at least three separate organizations, each with separate policies and purposes. Nevertheless, the process of integration would seem to have commenced, though it will take an organization like the Commission of the European Union to mobilize the necessary resources, power and operational example to change conventions that have prevailed for at least half a century.

One way would be to integrate disaster management as a component of environmental management. Natural disaster does, after all, result from natural hazards, and natural hazards do impinge upon the natural environment as well as emanate from it. Moreover, the effects upon the environment of the activities of society, have a direct bearing upon the impact of hazards upon society and its settlements, communities and activities.

The World Commission for Environment and Development (WCED, 1987) expressed their concern that failures to manage the environment and to sustain development threaten to overwhelm all countries. They were concerned for the achievement of a management system whereby environmental matters would be more successfully integrated with development:

In the past, responsibility for environmental matters has been placed in environmental ministries and institutions that often have had little or no control over destruction caused by agricultural, industrial, urban development, forestry, and transport policies and practices. Society has failed to give the responsibility for preventing environmental damage to the 'sectoral' ministries and agencies whose policies cause it, Thus our environmental management practices have focused largely upon after-the-fact repair of damage: *re*forestation, *re*claiming desert lands, *re*building urban environments, *re*storing natural habitats, and *re*habilitating wild lands. The ability to anticipate and prevent environmental damage will require that the ecological dimensions of policy be considered at the same time as the economic, trade, energy, agricultural, and other dimensions (p 39).

What closer parallel with disaster management could there be? Add only relief, rehabilitation and recovery!

The reduction of vulnerability should be made the responsibility of all sectors of development – those that in the past have perpetrated vulnerability. They are the same as those listed in the quotation above. Each one should be given guidance sufficient to enable it to understand the implications for socio-economic vulnerability that are incumbent within its activities – and within its intentions and objectives.

A procedure for vulnerability assessment (VA) could be made a social adjunct to environmental impact assessment (EIA). The history of EIA is similar to that of VA; it took time for the transition from an add-on technique to one which was eventually to realize its full value through application as a broad filter integrated within the development planning process, rather than one that was sometimes superimposed upon individual projects (Lewis and Carter, 1986). VA would be the human and social contribution that differentiates 'disasters' from other environmental impacts: it would be an indication of (generic) man's better perception of himself as a component of the natural environment, by which his own protection will be more assured, rather than his assumption of superiority, separation and distinction from it.

By initiatives of this kind, vulnerability could be made to reduce, and only by a reduction of vulnerability will disasters be made to reduce, now and in the longer term.

7

Equitable preventive development

Equitable practice

WHILE RECOGNIZING THAT the poorest are the most vulnerable to disasters, disaster 'prevention' and 'preparedness' have tended to construct their strategies regardless of whether or not these were effective for, and accessible to, the poorest – leaving 'disaster relief' to tend (or not) to their needs in the aftermath of emergencies.

Amelioration of the perceived causes of, and reasons for, vulnerability has to be applied equitably throughout all groups and all places – and at all times ('social equity' is a principal objective of Agenda 21: Rio, 1992). Responding only to events as they occur is likely to contribute further to social inequitability and socio-economic imbalance. A context of emergency is not conducive to ensuring equitable distribution of post-disaster assistance so as not to leave some 'groups' disadvantaged or externalized – or with their perception of being so; or with markets and prices being disrupted by the sudden availability of goods and materials. The avoidance of inequity and disruption requires careful planning and management that only normal circumstances can accommodate. This is the responsibility of development.

Once a strategy for the avoidance of inequity has been reached, post-disaster assistance, if required, will have a known and acceptable context within which to integrate.

That there are instances where relief consignments have been illicitly appropriated for the support of warring factions (Dommen, 1996), serves to underline the principle of equitability. The aim of war and conflict is to destabilize, or to exploit destabilization, for the purpose of achieving power – which is in itself, inequity. The achievement of equitable contexts serves to reduce the risk of war and conflict. Thus, management in the aftermath of natural disasters, of relief and reconstruction may, if the principle of equitability is not rigorously applied, contribute in the longer term not only to vulnerability to recurrent natural disasters, but to civil dissatisfaction, civil unrest and conflict. There have been numerous instances where political instability has ensued in the aftermath of natural disasters (see Chapter 3).

Economic interventions on their own can cause and aggravate social vulnerability (see also Chapter 4). Rehousing in Tonga after hurricane Isaac in 1982 required the financial participation by recipients of a quarter of the cost of a new dwelling which was to be provided for them. Many could not afford to participate, and others could afford to do so only after several years of saving. The poor who were not able to avail themselves of a new dwelling would be the most vulnerable to the next hurricane, to its physical impact and to highest proportional loss of the little they possess. Those who were rehoused, and those who were not, would be visibly and socially identifiable and separable. This is one example of how short-term policies – in this case for the management of rehousing after one disaster – may cause or exacerbate the further marginalization of those unable to participate. A policy which meant reconstruction for some has contributed to long-term vulnerability for others (Lewis, 1989b). In its manifestation next time, repeated and recurrent resources will be required if procedures are not modified and if obvious or preceived inequality, social dissatisfaction and possible social unrest are to be avoided.

Marginalization, with underdevelopment, is established as a significant source of vulnerability (Baird *et al.*, 1975). Therefore, processes which continue to contribute to existing marginalization and underdevelopment, or create new conditions of that kind, need themselves to be identified and avoided so that vulnerability may be reduced. Policies for financial participation in reconstruction, for example, perpetrate inequitable divisions between those who can and those who cannot participate, when all have the same need.

Equitable preventive development has the potential for reducing the socio-economic effects of natural disasters and of some conflicts, and for diminishing the likelihood of their recurrence or continuation. Accrued conditions of invidious deprivation and perceived disadvantage may lead to animosity and conflict, often under cover of more obvious ethnic differences. Where equitable development can be deployed to assist the establishment and growth of pervasive and comparative well-being, there may be more likelihood of ethnic differences being absorbed into an overall cohesive and gregarious social network.

Account has to be taken of local social and economic imbalances and of real or potential tensions (Macrae and Zwi, 1992). More often than not in the past, it has been usual to avoid such issues in a desire to remain neutral. Conflict and war aim at disempowerment of local groups and organizations. Local participation in recovery and development programmes requires and brings about local empowerment. Conflict reduction can also be a built-in objective of development, by explicit acknowledgement of tensions and in its consideration of the maintenance or achievement of equity and avoidance of differentials.

Decentralization and accessibility

Perceived need for the retention of power and control of finances and resources, the adoption of military styles of government, or simply the assumption of authoritarian and technocratic systems of management, are often manifest in centralized forms of government. Conversely, this has reduced the powers and responsibilities of local governments and of localized governmental sectors. This has usually lead to a diminution of participatory initiatives for localized social and economic development (Smith, 1982).

The functional local provision of socio-economic infrastructure, resources and services on the basis of population and need has little opportunity in centralized systems. Devolution and decentralization are required if smaller urban and rural populations are to be equitably and functionally accommodated and resourced. Political participation, credit and banking facilities, education and health services, agricultural extension services, are examples of services that may not be within the objectives or reach of centralized systems. Their absence or inaccessibility may have crucial implications for socio-economic vulnerability and for survival everywhere.

For example, although the rural participatory process briefly described in Chapter 3 facilitated identification by villagers of their micro-vulnerability, their interpretation of vulnerability focused on the impact of a natural hazard and not on needs for survival in the aftermath. Applications of rural participatory assessment provide the opportunity for the identification of local methods and needs for immediate survival, for example, whereas assessments of post-impact requirements are part of a macro-vulnerability.

Macro-vulnerability may best be identified not by immediately local partici-pation (Lewis, 1982a) just as it will need to be met by resources which may be outside the access, experience or perception of a community.

It might have been elicited that the nearest school was further away than young children were able to walk, or that the nearest clinic or health centre was further away than most people wanted to travel when in need of treatment – which had the effect of removing access to treatment altogether. Whether this kind of observation would have transpired in the Philippines is not known, but there are countless places where it would – given the positive opportunity of expression.

The centralization of health services in large general hospitals serves econ-omies of scale in provision and management. They are built to a size commensu-rate with the area served and the numbers of dependent population. Conversely, the larger the area, the larger the number of people who have greater distances to travel for access to the services they need.

On the other hand, decentralized health services, allocated according to dis-tribution of population, would each serve a smaller number of people. Accessi-bility to health services would be increased for more people as a result of distances being shorter. Services would be more accessible to more people and therefore would be more equitable.

In Bangladesh, as the result of rural participatory processes, it has been established that primary schools, serving a dual purpose also as shelters during cyclones, are preferred to be at a maximum distance of two-and-a-half kilo-metres for either small children walking to school or for adults with young or elderly dependants making their way to cyclone shelters.

The provision of social services for health and education, assessed on the basis of the users' need, and not only from the point of view of the provider, elicit a different strategy. The appropriateness of accessible and small-scale services becomes the greater when health services are interpreted as largely advisory and preventive and mostly to do with nutritional guidance, preventive medicine, accidents and minor ailments or with normal healthcare in say, pregnancy, childbirth and childcare.

The implications for vulnerability reduction are significant. A socio-economic interpretation of vulnerability reduction, inclusive of post-impact survival requirements, requires the localized provision of health services. Rather than it being expected or assumed that emergency hospitals would be 'flown in', would it not be more effective for the day-to-day achievement of a healthy population, for health services to be made normally accessible and not only (or not) in an emergency?

It can be argued of course that an earthquake, for example, would damage or destroy health services in the area affected, and render them inoperative. If this were the case, then with health services distributed according to population, there would be shorter distances to the nearest services that remained in operation. If the earthquake had destroyed the centralized services of a general hospital, then many more people would be deprived of those services – whether they were accessible to them or not!

Pervasive accessibility to resources made possible by the provision of numerous small-scale facilities distributed according to population, rather than centralized services, is a basis for equitable provision. Natural disaster damage to the service will be correspondingly decentralized and the vulnerability of the service will be reduced by less emphasis being placed upon major facilities serving a wide area and large numbers of people. Similarly, trained staff would be better deployed spread among the population, mobile and independent of technical resources. Disaster damage would occur to a larger number of smaller facilities. On the periphery of the area of damage, there will be undamaged facilities remaining in use, and not far away. People's vulnerability will correspondingly reduce by the greatly increased accessibility of health services, both to them and for them, and both in normal times and in emergencies. The same strategy for decentralization and accessibility applies to all public services.

Disaster preparedness strategy often includes the highlighting of buildings containing services of high strategic value. A form of strategic development for disaster reduction, on the other hand, would focus on reducing the number of buildings of high strategic value and on providing more installations and services that by their increased number and reduced size are strategically less critical.

There are some longer-term considerations to do with people's perception of their relationship with authorities and institutions. These can be said to parallel the relationship of children with a parent. The relationship can be one of love or one of fear – fear being of loss or abandonment (the relationship of fear produces a more tenacious relationship; the child in fear will cling closer to the mother). These immediate and eventual relationships impinge upon behaviour in later life, often with negative consequences (Zulueta, 1993).

In relationships between people and authorities or institutions, there will be a more positive relationship where the institution is represented locally and is accessible, creating an atmosphere of care and a sense of identity and belonging, than where the institution is remote and inaccessible. There are many regions where accessibility is so remote as to be out of the question; in these cases people are in a state comparable to abandonment and their behavioural response to authority or institutions, inclusive of their governments, is understandably negative – and often actively so.

140

Survival and recovery

In addition to the need to reduce or to avoid the primary effects of physical and personal damage, is the need that then ensues for human survival to continue. Provision for survival is synonymous with provision for vulnerability reduction; and provision for both is relevant to all kinds of emergencies and disasters. Vulnerability is not disaster specific, and neither are needs for survival. Means for survival in one instance are means for survival in another (Chapter 5).

Recovery after any kind of natural or other disaster depends upon the number of survivors, their capacity to continue to survive, and their condition before the catastrophe happened. The condition prevailing before a disaster – of a person, structure, community, or society – is of crucial significance to capacity to recover after loss, damage or destruction has been sustained (Haas *et al.*, 1977).

Survival is not only a humanitarian objective; survival is the prime prerequisite for recovery, and recovery is the crucial prerequisite for development to continue. Development must therefore integrate the means for survival.

It has been demonstrated in Bangladesh how social networks may facilitate survival in flood and after tropical cyclone where, without those networks, survival is less likely (Rahman, 1991). The networks facilitate social organization and the greater likelihood of a sharing of losses and of resources. Networks are largely the result of rural non-governmental organizations active in community development, or in womens' groups, for example.

One significant network is the Grameen Bank micro-credit system, which makes small credit available to individuals who have formed groups for the purpose among themselves. The Bank extends this facility to the poorest and to single parents, for example, who would be unable to avail themselves of conventional banking services and who are normally exploited by moneylenders. The Grameen Bank makes possible a degree of self-employment and self-sufficiency among the poor that would otherwise not be possible.

At times of disaster, normal repayments are suspended by the Bank, and those who have been able to save money have something to draw upon. The Bank is able to assist those who have no savings, with loans through the group system. Moreover, the members of networks formed by the Bank's system of credit availability are informally inclined to help each other in other matters.

In this way, survival is more assured for more people and recovery will take a shorter time due to resources available, and by self-sufficiency, and the products of self-employment will be interrupted for a shorter time. The Grameen Bank's small-credit system is socio-economic development in itself, and it helps further socio-economic development in the process.

People are the greatest natural resource; without people, there cannot be recovery. Survival is crucial to recovery and development. Recovery, as the rehabilitation of participation, will be enabled and facilitated by the previous creation in development of social structures for participation, and of accessible resources for basic needs. Having achieved and supported survival, these will become the resource for rehabilitation and reconstruction. Out of this process, locally identified and initiated recovery will ensue and feed further development.

Future recovery will follow as an expression and as the product of socially integrated sustainable development.

Vulnerability and sustainability

Development of basic needs provision for the reduction of vulnerability at local levels, in response to local perceptions of needs for survival, is directly compatible with the objectives of sustainability. In normal circumstances, development will continue beyond these basic requirements, into transportation, communications, energy generation, irrigation, marketing, small credit, for example – also compatible with community perceptions of needs to increase the range of available options through income generation. Institutional and organizational strengthening and learning for the articulation of appropriate planning initiatives and strategies will further assist the emergence of vulnerability reduction as an integral part of these processes.

Sustainable development meets the needs of the present without compromising the ability of future generations to meet their own needs (WCED, 1987). In the concept of 'needs' and the essential needs of the world's poor, and in responding to needs, the recognition of limitations upon technology, of social organization, and upon the environment, are key issues. Economic and physical development policies and strategies have a significant bearing upon sustainable development, their formulation and implementation being crucial to the success or failure of sustainable development (UNCED, 1993).

Sustainability is thus basically a social concept of people's needs and aspirations, and has to do with the relationship of those needs and aspirations with local environments (ComSec, 1991). Relationships between the satisfaction of 'needs' and perceptions of the environment are probably stronger at many community levels than they are in centralized or remote technological and economic considerations. Cultural linkages between society and environment are perhaps the strongest links of all. Such indigenous matters regarding agricultural diversification, building construction, naturally occurring foods and medicines and their preservation and storage, community systems and communications, are all relevant to local sustainable development. They are also relevant to the maintenance of vulnerability reduction (Thaman et al., 1979).

Social sustainability, in its own right, seeks to ensure the continuity and protection of social entities and in doing so '. . . to protect the vulnerable, respect social diversity, and ensure the fullest participation in decision making . . . (in short), to build up rather than to destroy, social capital' (World Bank, 1996).

In contexts inclusive of natural hazards, therefore, local environmental interrelationships which contribute to vulnerability reduction, contribute to the protection of the vulnerable and are components of social sustainability.

However, as more lives are saved by physical infrastructure (cyclone shelters; embankments; hazard-resistant construction), the socio-economic dimension of those lives will be exposed (Lewis, 1997). New questions will then emerge if they

142

are not addressed at the start: how will survivors stay alive? How will survival be transformed into recovery? What socio-economic inputs are required to facilitate self-reliance and sustainable recovery? Livelihoods have to be preserved in addition to the saving of lives. The saving of lives has itself to be made sustainable – by ensuring that lives saved can be self-sustaining. Save lives and sustain life.

Natural hazards are an integral part of the dynamic processes of environmental change and are thus an environmental and human ecological issue – because they arise out of natural phenomena; because they have impacts upon the environment of which man and communities are a part; and because the precondition of that environment has a bearing on vulnerability to natural disasters and survival in their aftermath. These factors impinge upon questions of social as well as environmental sustainability.

The extent and nature of impact conclude natural hazards to be a component of environmental sustainability; implications for human survival and recovery conclude natural hazards to be a component also of social sustainability (Berke and Beatley, 1997).

Were development to become environmentally and socially responsive, it would subscribe to disaster reduction; were development to contribute to vulnerability reduction, it would be more environmentally and socially responsive (Lewis, 1994b).

The impact of natural disasters upon development is often colossal, to which development must respond by being made more environmentally and socially conscious. Development has to be made inclusive of measures for survival, and consequent rehabilitation and recovery, in an understanding of the dynamics of vulnerability accretion. It would be appropriate for policies to emphasize that every action taken on account of one disaster must be designed and managed also to reduce vulnerability of the future.

In this way, vulnerability reduction itself would be socially and environmentally sustainable development. Development will be the more secure having taken into account environmental hazards and the risks of natural disasters. Appropriate development would also reduce the requirement, cost and long-term implications of emergency assistance (Lewis, 1980; UNRISD, 1993).

What is required, therefore, is adjustment of development programmes and projects to take account of people's needs so as to increase their options in and after disaster – so that they may better survive the initial impact and continue to survive and to recover in the short-, medium- and long-term aftermath. This would thus be social sustainability and vulnerability reduction as well.

To regard natural disasters themselves as 'the main obstacle to social sustainable development' is at once to separate disasters conceptually from development, and is simplistic. It is not sufficient for development only to preserve itself in a physical sense – and to call that 'sustainable'. Development must achieve its sustainability by the security it imparts to the lives and livelihoods of its beneficiaries.

It is interesting to observe how the principle of sustainability has been interpreted, first as an environmental issue in a physical sense, then as a socio-economic issue, and more recently as a social issue. What, it seems, has yet to come is sustainability as a political issue. It has been observed (above) how both complex emergencies and natural (simple) disasters have been caused, at least in part, in recent histories. Political decisions affecting the division, relocation and containment of populations have often led in their longer term to perceptions of inequality, deprivation, unrest and eventual conflict. This surely has reflected unsustainable policies of the past that have focused on the then present needs of governments and policy-makers but not on eventualities and needs of future generations – either at environmental, social or even political levels.

Further, it is essential to observe in this context how the ethics and policies of governments affect and influence the ethics and activities of their populations. The ethics, activities and behaviour of people are, at least in part, a direct consequence of the ethics and policies, as distinct from the regulations and laws, exercised by their governments.

8

Vulnerability reduction in development

A review

VULNERABILITY DEPENDS UPON conditions that are continually changing over periods of time and, therefore, the vulnerable condition itself also changes. Vulnerability is an ecological concept and is not static. Changes in social and economic conditions may bring about an increase, or a decrease, in vulnerability, even where the context of recurrent natural extremes remains constant. With improved identification of vulnerability, influences bringing about its increase or decrease can be further identified. By adjustment of these influences in a development context, processes identified as the causes of vulnerability can be impeded or stopped. The effects of hazards will be lessened over time and some disaster reduction will have been achieved. These measures do not belong to emergency aid, nor are they entirely within the boundaries of humanitarian assistance.

The impact of natural disasters (Lewis, 1990) and of war (Stewart, 1993) upon development has been colossal. Development has now to respond by identifying and incorporating prevailing risks of environmental hazards and of conflict, and by becoming more socio-environmentally conscious. Measures for survival, rehabilitation and recovery, need to be incorporated in an understanding of the dynamics of vulnerability accretion and, therefore, in such a way that provision for regions, communities, and social groups is perceived as equitable.

Development can be moulded to adapt to requirements. The degree by which populations and their activities can absorb, reduce, or change their vulnerability, accommodate hazards and achieve survival and recovery, is an expression of appropriate and prevailing political, social, cultural and economic conditions – the achievement of which can be made to be the objectives of development.

Vulnerability can be considered from two aspects – survival and post-survival (Lewis, 1987b). The first aspect relates to the initial impact of disaster, when vulnerability may be conditioned by location, density and distribution of population, age of people and of buildings, and technical and social capacity for resistance and protection. The second aspect relates to aftermath and the capacity of survivors to continue to survive in the longer term, for which will be required an effective culture or infrastructure of assistance, resources and social services (Finquelievich, 1987).

Disastrous extremes are extensions of a prevailing normal hazardousness. Small and local disasters cause conditions of vulnerability which contribute to subsequent larger ones. An *ad hoc* response only to extreme and rare disasters is inappropriate; it is the recurrence of small-scale frequent events that requires most attention, both in themselves as local experiences, and as creators of vulnerability. Further, the more frequent are events, the more normal they are – to the point where normality itself is the vulnerable condition. Attending to the small ones is preventive strategy for the large ones – and is more cost effective.

Thus, it is the more normal environmental condition, the enablement of survival in that condition, and the development of context and infrastructure at local levels that are required, not only global response to the more unusual, emotive or spectacular disasters (Lewis, 1987a).

The reduction of vulnerability, as the crucial path to disaster reduction, has to be focused not only upon 'protection' by technology in building construction, or in warnings and communications (e.g. UNCED, 1993), but in measures more to do with accessibility to social and material resources, and social initiative and participation as dimensions of cultural expression of traditional knowledge and norms. Disaster reduction has to be made inclusive of the enablement of human ecological adjustments in the activities of vulnerable people to maintain a resilience and self-reliance to counter the effects and implications of disasters, rather than only as technological resistance to the events and forces of environmental extremes.

Disasters – the monitor of development

Natural disasters are the monitors of development. In disasters are exposed the shortcomings, and certain strengths also, of preceding processes of change (see Case-study III). Whether these processes have been planned or whether they have been fortuitous, whether they have caused or exacerbated vulnerability, or whether they have reduced vulnerability, will be exposed in the manifestation of natural hazards.

What development has done or has not done, what it has failed to achieve and what it has excluded or ignored, is also exposed in disaster aftermath as 'the debt of development . . . disasters are unpaid bills' (IDNDR Technical Committee, 1998). The 'relief machine' moves into what development has preferred not to include. As a result, we are observing a new 'disaster imperialism' which has created for itself the opportunity for international interventions in the aftermath, an aftermath created by defaulting development processes.

That some of these processes began long before 'development' was a moral imperative, but were initiated by colonial imperatives and priorities, does not change this view. It simply means that the processes that have created vulnerability and inequity belong to the long term – and that they will require long term policies and processes for their amelioration. The perpetuation of short term policies and interventions will serve to attend to the victims as the symptoms of the system, but will also serve only to perpetuate the causes. This, as has been suggested above (see Chapter 7), is unsustainable. The power of the relief machine grows on the inadequacies of development. It is high time for this process to be counteracted and reversed.

Urban and rural balance

In cities, established and immutable as they are in their primary vulnerability (of location), their social services of water and food supplies, hospitals, clinics,

146

surgeries and dispensaries, and radio, newspapers and bulletin boards, all subscribe to continued survival in disaster aftermath. These indigenous services must be included in the objectives of development. These are the infrastructure without which survival will not be sustained and without which post-disaster relief cannot appropriately be integrated and directed.

The self-evident and colossal immutability of vulnerable cities and the attention these cities correspondingly command, justifiably suggest a special concern, but rural populations continue as an indicator of less-developed countries to be (still) the larger number who require similarly pervasive and accessible networks of social services for both normal and post-disaster conditions (Lewis, 1980). Without development on behalf of rural communities as well, development will be inequitable, and city migrant populations will continue to increase.

The significance of rural areas requiring rural programmes has been emphasized by the results of research in widely different social and economic conditions, as well as by the descriptions of conditions in Tonga and Algeria (see Part 1). Different reasons from different places strengthen the purpose of rural project recognition. It is an accepted characteristic of most developing countries that the larger proportions of population are rural. Attention to vulnerability reduction in rural and remote areas is imperative if natural disasters are not to continue to be a trigger for migration to towns and cities. Disaster reduction strategy has to be deployed on behalf of rural areas, where vulnerability may not be as physically obvious as it often is in urban contexts (see Case-study IV). Rural participation in data collection, for example, requires basic equipment, training, and technical assistance and could form the basis of replicated small projects for development assistance.

A pattern of development designed for vulnerability reduction requires an aggregation of numerous small-scale initiatives and interventions. It cannot provide all of the answers, but suggests a method of assessing and responding to hazards at local levels of experience. The constituents of such a pattern are nowhere near as exciting as disasters themselves, nor even as the humanitarian assistance that might follow them, and they may not be of great attraction in the public imagination!

Disaster definition depends upon an event that renders a population incapable of recovery without external assistance. If, therefore, by a kind of development appropriate to reduction of compound vulnerability, it became possible through development at local levels to enable communities to survive better and to recover without external assistance – then in this way disasters would be reduced.

Preventive development

Complex emergencies in many countries have created contexts in which conventional interpretations of post-disaster assistance and of peacetime development are inappropriate and unworkable. Desperate crisis-driven decision-making, often in competition rather than co-ordination, needs to be replaced by systems for interrelationship, integration and co-operation with a view to the medium and long term, as well as to immediate actions.

Relief and development have to become less distinct, and the distinctions between them best avoided, in what could be a practical relationship of each modified in a linkage expressing context and circumstances, needs and capabilities, time and change. Both conventional relief and development have to change in the face of circumstances that are repeatedly unconventional, in complex and continuing emergencies and cyclical relationships with disasters, towards the overriding objective of equitable vulnerability reduction. Reactive relief integrated with pro-active development could assist the achievement of self-reliance in the long term (Madrid, 1995).

By the forming of disaster response based upon analyses of vulnerability there comes a shift of emphasis from cause or causes identified as external, to those causes which are identifiable, understood and to some extent, controllable from within (Lewis, 1979a). Modification to those identified processes and conditions that create vulnerability, serves not only recurrent disasters over time, but serves disasters of all kinds, magnitudes and origins. Shortcomings and negative conditions, and certain strengths of indigenous systems, are quickly identified at local levels, as are those constraints and conditions over which local actions can have no influence.

The political dimensions of natural disaster vulnerability reduction are much the same as those which could lead to some kinds of conflict. With a more realistic understanding of processes that are the cause of vulnerability, there is the possibility that projects could be devised to respond both to a reduction in the possibility of conflict and of natural disasters. The principles of equitability, of socio-economic vulnerability and of self-reliance for survival have to be built into such programmes (see Chapter 9).

On the other hand, there is the risk that without an understanding of vulnerability and the need for its reduction in identifiable disaster-prone areas, policies may ensue for limitation of disbursements of development funding – in the erroneous belief that less will be lost when conflicts or disasters occur. Policies such as these, where they exist, would become the basis for perceived disadvantagement, dissatisfaction and inequitability, which are so often the basis for violence and conflict. They are fuel for the eventual fire.

By an understanding of vulnerability as a compound of social, economic and political interrelationships – of a survival infrastructure as well as a protective structure – 'preventive development' takes its place within a spectrum of measures for a corresponding spectrum of conditions. The dramatic and obvious physical effects of disasters are more appropriately understood as relatively superficial to the hidden, less overt, but longer-lasting social effects, and to those hidden processes that may bring about or exacerbate susceptibility and exposure.

By attending to the root causes of vulnerability, both post-disaster and pre-disaster socio-economic systems will be attended to. Post-disaster conditions will require less exogenous input, and normal 'between disaster' everyday existence will have been improved. Is this objective not also the objective of development?

There remains the need to continue to explore ways in which strategies for the reduction of vulnerability of human settlements can be incorporated more rigorously and more effectively into development programming, and as part of post-disaster rehabilitation and reconstruction (Lewis, 1983b). Vulnerability reduction depends upon institutional organization (see Chapter 6), development and co-ordination of otherwise separate sectors and activities, as much as it does upon technological developments within each sector. Disasters pervade all boundaries; interrelationships count for more than convenient separation of issues, sectors or regions. Holistic and systemic, not reductionist problem-solving and management are required for all environmental issues – including hazards.

The following are the essential ingredients of programmes for effective and identifiable reduction of vulnerability in development:

1. An understanding of the multi-sectoral and interrelated processes which create vulnerability, and of requirements for their national, regional and local identification and articulation.
2. Towards this identification, the accounting of disaster impacts needs to be made to be closely compatible with recognized and practised accounting systems for the monitoring and measurement of socio-economic development. In this way, the measurement of disasters will demonstrate both the impact of disasters upon development and, more meaningfully, the impact of development upon vulnerability.
3. Co-ordination of otherwise separate and possibly conflicting activities, and to establish and to assist the application of vulnerability assessments as part of all development proposals and initiatives.
4. Understanding of the socio-economic mechanisms and local impacts of hazards and disasters and to assist an equitable balance of responses across and between social groups, governmental and non-governmental activities, and between institutions and organizations.
5. Reintegration of the perception of vulnerability within those identified processes which could otherwise lead to its accretion; and to ensure that responses to disasters and conflicts address their socio-economic causes and contexts.
6. In the aftermath of disasters, to assess the shortcomings and strengths of established activities and development patterns and priorities, to use disasters as an opportunity for local project identification of medium- and long-term development needs, but to recognize that disasters are not a necessary prerequisite for this undertaking.

Implementation of policies for the integration of disaster reduction initiatives within development contexts requires a workable practice to be constructed as a bridge between prevailing pre-disaster conditions, post-disaster aftermath and continuing change. Whereas no single sectoral undertaking can achieve this wide-ranging task, development programming can and must do so. As it is, 'disaster reduction' will attend only to the symptoms and product of vulnerability, not to its causative processes. Demand for disaster reduction through vulnerability reduction may be increasing, but while the processes of the causes of vulnerability remain unaddressed, its effectiveness will remain limited to the extent of current understanding.

Project identification

Medium-term strategic programmes for dispersed and pervasive provision of resources for survival, for example, would be one appropriate development initiative. Another could be that all social and economic activities provide the opportunity as part of their normal processes of change and development, for the incorporation of consideration of environmental hazards and of needs for survival in their aftermath. Modifications could then be devised accordingly to perceived requirements. Incorporation into development programming of measures such as these is their only option; it is not feasible for this kind of approach to be implemented in disaster aftermath.

Long-term strategies for self-reliance will be of benefit to society in normal times as well as being its own insurance for survival and subsequent recovery after any kind of destruction or disaster. To reiterate – recovery after catastrophe and disaster is dependent upon the prevailing condition of society.

In the meantime, repetitious statements such as 'post-disaster assistance could be designed to ensure that it helps to maintain development momentum in the stricken area, as well as to relieve immediate suffering and deprivation' serve only to suggest that development is something separate to which all else aspires; whereas it might be that very development that has contributed to, or even caused, the disaster for which assistance has to be mobilized.

Instead, workable practices have to be constructed so as to form bridges between prevailing pre-disaster contexts, the post-disaster aftermath and continuing change and development for vulnerability reduction. Post-disaster project identification missions can be devised to do this, and should be made standard practice in parallel with missions for disaster investigations, post-disaster assistance, and evaluation.

Were they to be so, communications would likely be a prominent item for improvement. Although telecommunications are often the first to fail in most disasters, their rehabilitation for more robust operation becomes essential. Equally so are alternatives, such as all forms of land and sea transport, boat landing facilities, air strips, roads and bridges. Improved communications of all kinds will serve to increase everyday social and commercial intercourse; community self-reliance and facility for mutual support will be naturally reinforced – and in disaster, self-reliance is the prerequisite.

Equitable disaster reduction is a process of change in which disasters themselves take part, in large and also in small scale. To be integrated into this process, project implementation is likely to be widely muti-disciplinary, small scale, and numerously repetitive at local levels. Participatory development (ADB, 1998) is a prerequisite, but with awareness of influences upon vulnerability which themselves may not be local.

The imagery of possible ultimate catastrophe should not be made to preclude seemingly minor measures on behalf of the interim real condition. There is always an epicentre and a periphery to disasters of all kinds; but it cannot be known in advance where or to what extent these will be – be it earthquake, cyclone, flood or conflict. It follows, therefore, that by the initiation of measures which may be considered minor or to be of benefit only to less affected areas, that the periphery will as a result be limited in extent. In this way also, disasters will be reduced.

Reconstruction and implementation

Reconstruction has a metaphorical as well as a physical meaning. It is concerned with policies and systems as much as with the infrastructure that those systems both require and make possible. It is also concerned in a social as well as a physical sense. Not only is reconstruction concerned with the rebuilding of destroyed and damaged physical and social infrastructure, but with modification of the previously existing, where that is necessary for the achievement of equitability and the reduction of vulnerability.

Diffuse and pervasive programmes of rural development may be required to achieve or maintain reconstruction and development balance between urban centres affected by destruction, and rural areas, both those affected and those not affected. Account has to be taken of existing social, administrative, institutional and commercial systems as well as their physical infrastructure. Such a programme could be designed to increase considerably services of all kinds for rural areas.

In 1980, the city of El Asnam (subsequently renamed Chleff), together with its surrounding rural areas, was destroyed by earthquake (Lewis, 1982b). A programme for reconstruction of this kind was closely compatible with political and development policies in Algeria at that time – for agrarian revolution, autogestion (self-management) and localization (Sutton, 1978). It was also compatible with development for self-reliance as a counteraction to dependency and population movement to urban centres.

Not only was there more chance of a reduction of vulnerability to direct earthquake impact, but secondary effects of deprivation and homelessness would also be reduced by the increased availability of localized resources and infrastructure. As they become available, data from seismic micro-zoning could be incorporated into rural development planning.

A planning model was proposed (Lewis, 1981a) to be adjusted through the processes of preliminary application, incorporating data on population and its distribution, resources and activities, and on earthquake (and other) hazards. With appropriate socio-economic data, application of the model indicated areas of vulnerability by degree. With the incorporation of planned technical, economic and/or institutional measures (also identifiable from the model) vulnerability could be reduced, and earthquake disaster reduction could be an integral component of housing, settlement and development planning. Judgements could be applied where quantifiable data on hazard remained

151

unavailable but, at the same time, application of the model assisted the identification of specific information requirements. The model was applicable at *wilaya*, *daira* or *commune* level; application at *commune* level would more closely integrate planning for urban/rural balance.

Vulnerability is not a static condition; it is dynamic, evolutionary and accretive (Westgate and O'Keefe, 1976; Lewis, 1984a). Both planned and spontaneous population movement and housing reconstruction after the 1980 earthquake (and not least, the earthquake itself) had conditioned vulnerability to the next inevitable occurrence.

Without co-ordinated and integrated policies and programmes of this kind, separate or sectoral policies and decision-making may induce or condone more dwellings in more hazardous areas. At the same time, the housing construction sector is being encouraged to adopt improved construction techniques. Only to institute codes and regulations for improved resistant construction, or to establish warning systems, denies prevailing vulnerability; more dangerously it suggests the accomplishment of 'solutions' and a safety that may not in fact exist. Holistic policy and decision-making frameworks are necessary to prevent these efforts in improved building construction being swept away in the next landslide, eruption, flood or storm surge.

The physical disciplines of civil engineering and architecture often dominate in the organization of implementation projects for disaster reduction. In addition to their acknowledged pedigree of social concern, their predominant involvement may reflect a formerly conventional and restricted view on behalf of initiators of projects that disasters are mainly about damaged buildings and physical infrastructure. With the onset of a possibly more informed and enlightened approach, sociologists and environmentalists, for example, are now incorporated into the implementation process. Overall management, however, remains with the physical disciplines.

This procedure might appear to reflect an advancement in project management, but between engineering and sociology, for example, there are some marked differences in approach. One crucial practical difference is the time required for working practices. Whereas engineering excels in the achievement of shorter implementation times, sociologists often require the involvement of field teams in rapid rural appraisal or rural participation programmes. 'Rapid' is a comparative term, and time required often cannot be reduced without sacrifices to the end result – and often not at all if requirements of the process are to be met adequately. The result is a significant discrepancy in the amount of time required by each discipline to the extent that, to add components for 'women-in-development' or 'rural participation' on to civil engineering projects, often means that these must commence some considerable time before the start of civil engineering contracts – and within most technical assistance contracts, they cannot!

The question can surely be asked, what would be the result if civil engineering components were injected into contracts managed overall by sociologists or environmentalists? A different kind of project would result.

152

9

Strategic development for vulnerability reduction

THE CONVENTIONAL LINKAGE between disasters and development starts with a natural disaster and considers what needs doing so that subsequent disasters are less severe. There is nothing particularly wrong with this in itself except that, while this disaster-centric and disaster-specific process is continuing, other projects may be being undertaken which are not part of those post-disaster initiatives. While disaster-specific initiatives are designed to make things better next time and in the longer term, other activities undertaken at the same time could inadvertently subscribe to making things worse. There are cases where 'development' has been the cause of disaster (Chapter 4); and there are countless examples of 'development' that has caused disasters to be more, not less severe.

Overall strategies, inclusive of all sectors across the development spectrum, are required as a matter of policy by all governments and development agencies for the identification, programming and local implementation of development for vulnerability reduction. Within such strategies will be exercised:

- programme objectives
- project identification
- macro- and micro-vulnerability assessments
- project planning and programming by all sectors
- co-ordination
- ante- and post-evaluation
- post-disaster and post-conflict analyses

and they will assess all development inputs, projects and programmes with regard to their effect upon equitability and vulnerability.

An important aspect of assessment is likely to be the degree of decentralization and consequent local accessibility to resources and services. Decentralization facilitates the adaptation of national programmes within various local conditions and changing or unanticipated circumstances, and in local contexts of cultural, socio-economic and political factors. Decentralization policies also serve to enhance opportunities for local people and organizations to participate in decisions and their implementation, which have a bearing upon their lives and well-being. Decentralization is a pre-requisite for social sustainability in development.

Strategic development for vulnerability reduction will consider the social and economic, as well as physical infrastructural components and dimensions. Physical and infrastructural development may be required to save lives (e.g. embankments, cyclone shelters and other buildings) but as more lives are saved, livelihoods will also need to be preserved for sustained survival and recovery. This is the crux of the distinction between physical and socio-economic vulnerability.

Multiple and multi-sectoral small-scale micro-projects will better achieve successful local integration, and appropriate infrastructure will be selected for its potential for socio-economic regeneration. These could include, for example:

- *Transportation and marketing*: the provision of bridges, roads and pathways, boat and light-plane landing facilities, essential for commercial and marketing activities (for an improved quality of life and well-being – the essential prerequisite for recovery)
- *Socio-economic activities during floods*: the construction of bridges, sluices and culverts
- *Small industries*: rural energy generation and electrification
- *Agriculture*: projects for subsistence agriculture and small-scale irrigation
- *Health and education*: clinics and health centres; primary schools.

Social development includes programmes for community development, integration of non-governmental organizations, women-in-development, disaster preparedness training, preventive medicine and health services, maternity and child care, nutrition monitoring and small-credit schemes (Table 1).

It might also be useful to indicate what kinds of development are inappropriate to vulnerability reduction and therefore should be avoided. Examples are: development for growth (only); projects requiring or causing migration and/or resettlement; projects resulting in loss of lifestyle and traditional resources; and inequitable development or allocation of resources which could cause or exacerbate migration and cultural or ethnic divisions.

The development objectives indicated in Table 1 have to be achieved simultaneously and in combination, not sequentially or as one-off, piecemeal or *ad hoc* alternatives. As objectives towards poverty reduction, the objectives of vulnerability reduction fit well and are significantly represented, especially within overall policies for sustainable development (DfID, 1997; ODI, 1998: 2). Vulnerability reduction and poverty reduction are thus largely synonymous (Table 2) but although poverty reduction as an overall objective will contribute to vulnerability reduction, it is not the same thing. The processes of vulnerability accretion require identification and countering especially within contexts of poverty.

The link between the objectives of 'economic growth' and 'development at local levels', it has come to be realized, is not automatic; each is a different target

Table 1 Physical and social development for vulnerability reduction (Lewis, 1996)

PHYSICAL DEVELOPMENT	(for) SOCIAL DEVELOPMENT
• roads and pathways, bridges	• community development
• sluices and culverts	• NGO integration
• rural electrification/energy generation	• women-in-development
• health and education	• disaster preparedness training
• markets	• medicine/health services
• food storage	• maternity and child care
• agriculture	• nutrition monitoring
• small industries	• small credit accessibility

Table 2 Basic needs for poverty reduction

Basic needs	Vulnerability reduction
Enough to eat	Yes
Clean water	Yes
A livelihood	Yes (logistics/marketing small-credit)
A home	Yes (hazard resistant)
An education	Yes (information/training)
Health care	Yes
A safe environment	Yes
Protection from violence and natural hazards	Yes
Equality of opportunity	Yes (equitability)
A say in the future	Yes (development strategy/community development)

(Source: OXFAM [adapted in Lewis, 1996])

requiring different policies and operations. Of all the 'styles of development' that have evolved, none has adequately addressed the issue of *survival* in the context of natural or other hazards.

For this crucial objective, what is required is indigenous provision for basic needs in a human–ecological and sustainable approach to development. It is not sufficient for development to call itself sustainable, simply because it has incorporated some attention to a potential for natural hazards. It is development on behalf of its beneficiaries, in each of its parts and constituents, that has to achieve sustainability, if vulnerability is to be permanently reduced.

Socially sustainable development inclusive of the poorest, and of underprivileged sectors of the poorest, will further the removal of dependency – a significant cause of vulnerability. This will require local consultation, initiative and participation, with benefits in the short term as well as a longer-term future. Where there is prevailing natural hazardousness, there is an inherent urgency for action.

Basic needs for human survival and its continuation are:

- food and cooking facilities
- potable water
- shelter and warmth
- treatment of injuries, health, welfare
- information: the occurrence, what action to take and where to go for assistance (Lewis, 1987c).

Development for basic needs provision for the removal of dependency and vulnerability at local levels, and in response to local perceptions of needs for survival, is directly compatible with the objectives of social sustainability. In

155

normal circumstances, development will continue beyond these basic requirements into, for example:

- transportation and communications
- energy generation
- irrigation
- marketing
- small credit.

These require designing and programming to be compatible with community
perceptions of needs and to increase the range of available options through
income generation. Institutional and organizational strengthening and learning
for the articulation of appropriate development initiatives and strategies, will
further assist the emergence of vulnerability reduction as an integral part of
these processes.

Development as an appropriate and positive process has to be made inclusive
of counter-responses to the negative processes of vulnerability which may have
become ingrained at national and at local levels. Disaster reduction through
vulnerability reduction is a long-term process and will require appropriate and
commensurate planning and programming. In time, given appropriate policies
for the long term, established negative processes will be eroded and will slowly be
made to disappear. In the meantime, one-off short-term interventions are unlikely to be satisfactory in the longer term, in spite of there being no end to
demand and apparent need.

Development for vulnerability reduction must therefore be:

- informed, otherwise it may be counter-productive
- multi-sectoral
- positive (no 'development reduction in disaster prone areas'!)

bearing in mind that:

- poverty reduction and vulnerability reduction are closely compatible, but are
  not the same
- development for disaster reduction does not need a disaster to start it off.

A pattern for development designed for vulnerability reduction requires programming and implementing as a broad framework outlined above. In addition,
the following are the principal concerns and observations expressed in the
preceding chapters and Case-studies.

Insider assessments

Programmes are to be initiated not upon global aggregates of impact and loss,
nor solely upon in-country information received via the international media, but
upon intra-national comparisons of proportional impact, and 'insider' descriptions of circumstances and experience at national and local levels. These will be
the basis for insider motivation and participation, and the consequent opportunity for small-scale projects, both rural and urban.

Interpretations and assessments of 'disaster' by insiders are better able to convey local experience. Remote international comparisons cannot always do this (for example, drought in one place with its special circumstances of need may not be considered a drought in another with more rigorous rainfall regimes).

Proportional impacts upon places reflect the comparative scale of disasters. The greatest scale of impact is invariably upon the smallest places, smallest administrative areas and/or smallest units.

Small countries, small places and small units should not be disregarded in favour of larger (and more accessible) places where disasters are inevitably more impressive.

Recurrence

An awareness of all potential hazards at one place is essential for the experience of vulnerability and its interrelated processes to be identified and amalgamated.

The concept of the potential for recurrence of disasters of the same or different 'kinds' and in the same places is crucial for its bearing upon understanding of:

- the prevailing need to have regard to the implications of actions in the aftermath of one disaster as the context of the next
- the reality of natural hazards and vulnerability
- the dangers of a 'one-off' approach to single disasters
- the potential impact of one disaster upon the aftermath of an earlier one – of the same or different 'kind'
- the reality of hazard in relation to other development activities
- the impact of hazards themselves upon vulnerability (to the next)
- recovery from one disaster necessarily postponing development on behalf of another.

Assumptions about the size of a problem and extent of a condition in disaster aftermath can be counter-productive.

Survival, equitability, sustainability – and interrelatedness

Measures are required for strengthening self-reliance in development so as to achieve immediate and longer-term survival and recovery, as well as measures for emergency assistance and impact resistance.

Such measures should be implemented on a basis of equitability and made applicable to all groups, levels and sectors, and to all local activities and undertakings.

Equitable programming should be inclusive of rural/urban balance and should have in mind that socio-economic vulnerability is not hazard specific.

Rural/urban programmes should be aware that:

- areas of highest population density are not always the areas of highest losses
- rural damage and loss can be greater than urban damage and loss, though less dramatic, less impressive and less accessible

- rural populations in aggregate may be larger than those in urban areas
- lower density areas may suffer greater losses than higher density areas even though they may be in areas of perceived lower risk, i.e.: inland of coastal ports.

Sustainability is environmental and social. Policies are required for perpetration and enablement by which social and environmental sustainability will be achieved. Social sustainability may be achieved through environmental sustainability – and vice versa. All are synonymous with, and contribute to, vulnerability reduction.

Connectivity and interrelationships between issues and conditions is as significant to vulnerability reduction as the implementation of mono-sectoral undertakings.

Afforestation and tree planting are invaluable for numerous reasons and could be a part of many and various projects, being advantageous in context of floods, landslides, erosion, storm-surges, storms and high winds, and relevant as a possible (sustained) source of fuel and building construction material.

Local consultation, initiative and participation are crucial for the identification of interrelationships and consequences of projects, as well as for their local integration.

Dependency in all its forms must be avoided and/or removed.

Populations

Some crucial aspects are:

- population movement; spontaneous, coerced or forced
- permanent refugee and seasonal migrant communities and settlements
- the effect of large-scale activities upon those of others
- conflict; both direct and indirect effects and the prevailing need to reduce the perceived need for conflict by the implementation of equitable preventive development
- population density may be the first, as the most easily accessible, indicator of vulnerability, but population density in itself is not sufficient. Its absolute application would mean that all small populations are less vulnerable – which is nonsensical.

Social infrastructure (*infrastructure for socio-economic purposes*)

Emphasis on physical development and upon technology (which there surely is) has to be equated with compatible, integrated and simultaneously implemented social programmes so as to mirror local combinations and interrelationships of physical and social events, conditions, contexts and circumstances.

The implementation of social programmes requires appropriate and adequate time, whereas perceived achievement in physical projects is execution in minimum time. Therefore, for social programmes to ride the backs of the physical, may not be practical.

One objective for the integration of social programmes might be the regeneration of local historic and cultural coping strategies, methods and response systems.

Another might be for the provision of basic resources and services for remote populations – however small; in health, primary and secondary education, banking and small credit and transportation – noting the importance of communications between remote areas and settlements, as well as to and from remote areas and their capitals.

Local vulnerability requires local programmes of local significance; small-scale and repetitive projects are likely to be most effective.

Physical and other infrastructure

Encourage, enable and inform hazard-resistant domestic construction, inclusive of careful attention to simple reinforced concrete and of masonry construction in earthquake, cyclone and flood-prone locations, having in mind that:

- buildings protect each other; in clusters, groups and integrated with other features, e.g. new and existing trees.
- maintenance is a crucial aspect for the reduction of physical vulnerability – of buildings and infrastructure, and also of information/communications systems and all knowledge and expertise.

Emergency assistance

Beware the negative consequences of emergency assistance, for example:

- introduction or exacerbation of dependency
- disruption of local markets and prices.

Be aware of acceptance as a cultural response to gifts – judgement and initiative have to be with the donor.

The impact of post-disaster relief on indigenous administrations is likely to reduce local capacity to identify, assess and to adjust those structural weaknesses that exacerbate vulnerability.

Policy and administration

Existing and new development projects and programmes require the application of vulnerability assessments.

Administrative separation of the control of processes of vulnerability impedes their identification and that of their interrelatedness. Adopted and inherited administration systems which separate and divide, may require reassessment and reorganization.

Accurate and consistent accounting of disaster incidence is required to be maintained and commensurate with regular statistical records of socio-economic development progress.

Requirements for financial participation in reconstruction need careful consideration for their longer-term consequences as well as their immediate benefits – the exclusion of those who cannot afford financial participation may exacerbate their future vulnerability.

The implementation of strategies for vulnerability reduction are likely not to be contained within a 'disasters' unit, office or department, but be undertaken by a variety of 'desks' within existing development programming and management functions. These strategies could be of environmental relevance and are likely to be wide ranging and numerous and to require co-ordination.

What is wrong is not that things are not being done – though more needs to be done – but that things are being done mono-sectoraly under programmes to do with the providing sector – not to do with the potential recipient or a generic need.

Appropriate development for vulnerability reduction and equitable preventive development is being displaced by a new 'disaster imperialism'. At the same time, development would rather not be part of the disaster ethic, for fear of being proved to have been doing the wrong thing. Natural disasters are, after all, the monitor as well as the destroyer of development.

Those 'disasters sectors' that there are, are obliged to devise or retain lesser aspects which they are able to keep to themselves, because they are not part of the remit of any other sector. It is no wonder that 'disasters' remains separated from development and from all those causative processes for which it has to find counter-measures:

> . . .an extreme physical event can inflict a major disaster without prompting significant changes in hazard-management programmes or policies. This familiar circumstance is exemplified by the continuing plight of many people in third-world countries, where poverty and lack of resources often impede effective hazard management. . . hazard components are embedded in a variety of contexts. . . where societies are beset by different types of crises, it is desirable to design new public policies in light of their interrelated character . . . [these] will benefit from close collaboration among many types of people. They should not be confined only to scientists and managers whose work focuses specifically on hazards. Instead they should include the broad spectrum of organizations and interest groups that affect (and are affected by) the many contexts of hazard (Mitchell *et al.*, 1989).

In the approaches to the commencement of the International Decade for Natural Disaster Reduction, it was evident that there existed a prevailing preoccupation with regard to the magnitude of disasters and with technological resistance to hazards. The following summarized the situation then – and is not irrelevant now:

> Emphasis on, or the influence of, disaster magnitude, global comparisons of vulnerability, and exclusive focus on 'rapid onset' disasters, inadvertently obscures some key issues:

> o Expression of disaster magnitude reflects a privileged comparative view which tends to exclude disasters of a lesser degree, even though locally these can represent catastrophic impact. Magnitude thus tends to obscure the concept of local experience, understanding and context.

> o Vulnerability is similarly and essentially a local condition, and its under-standing invariably ensues out of local experience and analysis. Though vulnerability pervades in a global sense, its understanding is not assisted by generalized 'globalization'.

160

o 'Rapid onset' natural disasters may usefully delimit an enormous task identified for the IDNDR; but it is surely academic and, in current conditions, it is artificial. How can an *International* Decade ignore famine conditions now affecting at least 15 countries in Africa with a total area half again that of the USA? Furthermore, what is happening in Africa now is likely to continue for a very long time; it is merely a portent of what is to come.

o Inclusion of 'slow onset' disasters, whilst enabling the Decade to be more realistic, would at the same time facilitate understanding of those local conditions which contribute – the accretive factors of local susceptibility to disaster which collectively cause a condition of vulnerability. All disasters are slow onset when realistically and locally related to conditions of susceptibility.

o Local susceptibility relates to both initial damage and to needs for survival in the aftermath; because slow onset disasters display less of a distinction between onset, impact, and aftermath, they relate not only to damage to physical structures, but also to capacity of institutional systems, infrastructures, and management to survive and be the source at local levels of the survival of others. Survival is the essential prerequisite of recovery.

o The application of scientific and engineering knowledge is a partial response only to the pervasive, indigenous and accretive conditions that constitute susceptibility – the root of vulnerability. Scientific knowledge must surely include a greater measure of the sciences of sociology, anthropology, and geography, for example, so eminently represented in North America and which have been internationally at the forefront of field analysis and understanding of natural disaster impacts.

o Finally, in addition to the application of known scientific and engineering principles, where these are relevant and useful, there must surely be the readiness to analyse and more fully understand at local levels how contexts may have contributed to the damages that have ensued. Social and economic disruption caused by disasters cannot be adequately assessed or understood without knowledge of conditions that prevailed before the disaster impact. The same is true of indigenous systems for coping in disaster aftermath and survival.

o There could, in fact, be danger in the application of science and technology for disaster reduction. A focus on building construction, for example, may inadvertently exacerbate disasters by shifting the zone of effect or by ignoring a zone of greater influence. There are numerous instances where one hand is paying increasingly for preparedness against what the other hand is causing. Technology serves often to exclude by its applications: those who are not able to partake of it become successively more marginalized, more deprived, and thus even more vulnerable.

o Many large, and small, disasters recur regularly in countries where the range of choices of what to do about them is severely limited. Choice made possible by international intervention can lead to only temporary and often superficial 'solutions'. Disasters are caused as much by prevailing social and economic conditions as by severe natural extremes. It is the prevailing social and economic contexts that must be the target for our resources, not only the rare or unique disasters of large magnitude.

o Disaster reduction is best achieved from within and by support of indigenous knowledge, rather than by external applications. We should not so much be seeking to *confront* natural disasters, as to explore them on their own ground (Lewis, 1988a).

10

The next ten years and more

As THE INTERNATIONAL Decade for Natural Disaster Reduction comes to its end, two prevailing conclusions can surely be drawn. The first and paramount is that 'natural disasters' have become compounded, on the one hand by a more sophisticated understanding of their causes and, on the other, by pressure of events. 'Complex disasters' now reflect a realism of interactions on the ground between environmental hazards, disadvantagement, political instability, conflict and population displacement.

The term 'complex' suggests that another kind of natural disaster is/was by comparison 'simple', but the real issue is that our responses to all 'natural disasters' and their causes has been relatively simplistic, selective and partial – and in many quarters still is as a result of the overriding objectives of convenient and manageable institutional, ideological or political standpoints.

The second is that, at the same time, it has gradually become more clearly established in some other sectors that the causes of natural disasters are, or could be, within the capacity of societies for their amelioration and reduction. This gradual and partial shift has nevertheless probably contributed to the distinction that prevails between the 'relief' and 'development' lobbies – a distinction that has not yet been made to go away. The impact of the one upon the other, in monetary terms, is massive, a fact not unnoticed by the development bodies and which has led to a more critical attitude and desire for more control of humanitarian assistance (ODI, 1998: 1). Development budgets have nevertheless been greatly diminished by the growth of the need for relief – as a direct result of pressure of events if not of changes in policies.

The impact of the one upon the other in conceptual terms, however, may be more beneficial. That conflict is more obviously not a 'natural' event, may have had a bearing on a wider realization that natural disasters are, at least in part, man-made – especially in a context where the same humanitarian organizations are involved in both, where linkages may be newly explored (Lewis, 1988a), or where new debate in fresh forums may enlighten wider audiences (UNRISD, 1993).

A report of discussions on conflict and development co-operation (OECD, 1997) has a definition of emergency relief, rehabilitation and development operations which states 'these . . . forms of assistance [are best not] classified according to any presumed logical or chronological sequence leading from relief to development, which is rarely seen in reality. Failure to ensure that these operations are structured to be mutually reinforcing, however, can result in their becoming mutually undermining'.

The debate between and about the two approaches and their contexts continues (Alexander, 1997; McEntire, 1998), but reasoned analysis and discussion between them is made more difficult because contexts for implementation are of

162

such obvious political division. For there to be a merging of the two approaches on the basis of needs on the ground, is the prime prerequisite for change.

Disasters are at least 'the monitors of development' (see Case-study II) or at most, 'lack of prevention [in the past] is the debt of development, and disasters are unpaid bills' (IDNDR Technical Committee, 1998). There is indeed the need for an 'active international platform' to initiate the commitment, strength of purpose, resources, expertise and energy to merge paliative with preventive purpose into the next century.

References

ADB (1998): *Special Issue on Participatory Development ADB Review* Asian Development Bank. Manila.

Alexander, David (1997): The Study of Natural Disasters, 1977–1997: Some reflections on a changing field of knowledge *Disasters* 21/4 Blackwell. Oxford. pp284–304.

Angenheister, G (1921): A Study of Pacific Earthquakes *New Zealand Journal of Science and Technology* IV/5 November. Wellington.

Baines, Graham; McLean, Roger (1976): Sequential studies of hurricane deposit evolution at Funafuti Atoll *Marine Geology* 21 Elsevier.

Baird, A; O'Keefe, P O; Westgate, K; Wisner, B (1975): *Towards an Explanation and Reduction of Disaster Proneness* Occasional Paper No 11 Disaster Research Unit. University of Bradford.

Ball, D (1973): *Funafuti, Ellice Islands; Physical Development Plan* BRE/Department of the Environment. Garston, Watford.

Bayliss-Smith, T P (1977): Hurricane Val in North Lakeba: Population and Environment Project in the Eastern Islands of Fiji *Island Reports 1. Man and the Biosphere Programme.* UNESCO/UNFPA. Canberra.

Berke, Philip R; Beatley, Timothy (1997): *After the Hurricane: Linking recovery to sustainable development in the Caribbean* Johns Hopkins University Press. Baltimore.

Blaikie, Piers; Cannon, Terry; Davis, Ian; Wisner, Ben (1994): *At Risk: Natural hazards, people's vulnerability and disasters* Routledge. London.

Brindze, Ruth (1973): *Hurricanes; Monster storms from the sea* Atheneum. New York.

Brown, Barbara J (1979): *Disaster Preparedness and the United Nations: Advance planning for disaster relief.* Pergamon/UNITAR (and previous unpublished version).

Burton, Ian; Kates, Robert; White, Gilbert F (1968): The human ecology of extreme geophysical events *Natural Hazards Research Working Paper No 1.* Department of Geography, University of Toronto.

Campbell, J R (1997): Hurricanes in Kabara: Population and Environment Project in the eastern islands of Fiji *Island Reports 1. Man and the Biosphere Programme.* UNESCO/UNFPA. Canberra.

Cernea, Michael M (1996a) *Bridging the Research Divide: Studying refugees and development oustees* World Bank Reprint Series Number 481.

Cernea, Michael M (1996b) *The Risks and Reconstruction Model for Resettling Displaced Populations* Keynote Address Refugee Studies Programme, Oxford. September.

Clarke, Roger (1997): The challenge of getting donors involved in disaster mitigation In *Mitigating the Millennium: Proceedings of a seminar on community participation and impact measurement in disaster preparedness and mitigation programmes* (Jane Scobie; ed) IT Publications. London.

Commission of the European Communities (1996a): The European Union and the issue of Conflicts in Africa: Peace building, conflict prevention and beyond *Communication SEC(96) 332.* Brussels.

Commission of the European Communities (1996b): Linking relief, rehabilitation and development (LRRD) *Communication COM(96) 153 final.* Brussels.

Commonwealth Secretariat (1985): *Vulnerability: Small states in the global society.* Report of a Commonwealth Consultative Group. Commonwealth Secretariat, London.

Commonwealth Secretariat (1991): *Sustainable Development: An imperative for environmental protection* Commonwealth Secretariat London.

Commonwealth Secretariat (1997): *The Future for Small States: Overcoming vulnerability* Commonwealth Secretariat. London.

Cuny, Frederick C (1983): *Disasters and Development* Oxford University Press, Oxford.

Descloitres, R; Descloitres, C; Reverdy, J C (1973): Urban organisation and social structure in Algeria In *Man, State and Society in the Contemporary Maghrib* Zartman, I W (ed) Chapter 36 Pall Mall Press.

DfID (1997): *Eliminating World Poverty: A challenge for the 21ˢᵗ century*: A Summary. Department for International Development. London.

Dommen, Edward (1996): Humanitarian Aid in the Cycle of Armed Conflict *DHA News* June/August.

ECLAC (1985): *Damage Caused by the Mexican Earthquake and its Repercussions upon the Country's Economy* United Nations Economic Commission for Latin America and the Caribbean.

ECLAC (1987): *The Natural Disaster of March 1987 in Ecuador and its Impact on Social and Economic Development* United Nations Economic Commission for Latin America and the Caribbean.

ECLAC (1988): *Damage Caused by Hurricane Joan in Nicaragua: Its effects on economic development and living conditions, and rehabilitation and reconstruction needs* United Nations Economic Commission for Latin America and the Caribbean.

The Economist (1996): 22 June London. pp 49 and 52.

van Essche, Ludvic (1986): *Planning and Management of Disaster Risks in Urban and Metropolitan Regions* International Seminar on Regional Development for Disaster Prevention UNDRO Geneva.

Finquelievich, Susana (1987): Interactions of social actors in survival strategies; the case of the urban poor in Latin America *ifda dosier* 59 May/June.

Foucault, Michel (1971): *Madness and Civilisation* Tavistock Publications. London.

Fournier d'Albe, E M (1976): *Contour Mapping of Seismic Areas by Numerical Filtering and Geological Implications* Intergovernmental Conference on the Assessment and Mitigation of Earthquake Risk, Paris.

Golec, J (1980): *Aftermath of Disaster; The Teton dam break* PhD dissertation. Disaster Research Centre, Ohio State University.

Gribbin, John and Mary (1997): Weatherwatch *The Guardian* 8 August.

Grimble, Arthur (1957): *Return to the Islands* John Murray. London.

Haas, J Eugene; Kates, Robert W; Bowden, Martyn J (1977): *Reconstruction Following Disaster* The MIT Press.

Hall, Nick (1996): Coping with typhoons in the Philippines: Builders and farmers tell their story *BASIN News* No 12 August. Building Advisory Service and Information Network. St Gallen.

Hall, Nick (1997): Incorporating Local Level Mitigation Strategies into National and International Disaster Response In *Mitigating the Millennium: Proceedings of a seminar on community participation and impact measurement in disaster preparedness and mitigation programmes* (Jane Scobie; ed) IT Publications. London.

Hewitt, Kenneth (1983): The idea of calamity in a technocratic age In *Interpretations of calamity* (Hewitt, K: ed) Allen & Unwin.

Hewitt, Kenneth (1997): *Regions of Risk: A geographical introduction to disasters* Longman.

Hewitt, Kenneth; Burton, Ian (1971): *The Hazardousness of a Place: A regional ecology of damaging events* University of Toronto.

Hossain, Hameeda; Dodge, Cole P; Abed, F H (1992): *From Crisis to Development: Coping with disasters in Bangladesh* University Press. Dhaka.

Houghton, John (1997): *Global Warming: The complete briefing* (Second Edition) Cambridge University Press.

Huggins, Rich (1998): The Year 2000 Hazard: A hazards research and applications bonanza *Natural Hazards Observer* XXIII/2 Natural Hazards Research and Applications Information Centre, University of Colorado.

IDNDR Technical Committee (1998): *Washington Declaration* June. Geneva.

IDS (1994): Linking Relief and Development *IDS Bulletin 25/4*. Institute for Development Studies, University of Sussex.

IPCC (1990): *Climate Change: The IPCC scientific assessment* Intergovernmental Panel on Climate Change. WMO/UNEP Cambridge University Press.

Islam, Sirajul (ed: 1992): *History of Bangladesh 1704–1971: Political history* Asiatic Society of Bangladesh. Dhaka.

Jackson; Commander Sir Robert (1986): Disasters and the United Nations; International operations, science and politics *Interdisciplinary Science Reviews* 11/4 pp326–345.

Jeffery, Susan E (1981a): Our Usual Landslide: Ubiquitous hazard and socio-economic causes of natural disaster in Nusa Tengara Timur, Indonesia *Natural Hazards Research Working Papers No. 40* University of Colorado.

Jeffery, Susan E (1981b): 'The creation of vulnerable populations' Unpublished manuscript *Project for Natural Disaster Vulnerability Analysis* Centre for Development Studies, University of Bath. Mimeo 41 pp.

Jeffery, Susan E (1982): The creation of vulnerability to natural disaster: Case studies from the Dominican Republic *Disasters* 6:1 pp38–43.

Kates, Robert W (1970): Natural Hazard in Human Ecological Perspective: Hypotheses and models *Natural Hazards Working Paper No 14*. University of Toronto.

Kates, Robert W (1980): Disaster Reduction: Links between disaster and development In *Making the most of the least: alternative ways to development* (Berry, L and Kates, R W: eds). Holmes & Mier.

Kavaliku, S L (1974): *Letter to the New Zealand High Commissioner* (MW.12/6/2) 26 November. Nuku'alofa, Tonga.

Keane, Shake (1979): *The Volcano Suite; a series of five poems* St Vincent.

Kovats, Sari (1996): Climate change and human health *Tiempo* Issue 21 September. IIED/UEA.

Kuroiwa, Julio (1982): *Studies on the prevention of earthquake disasters and their application in urban planning in Peru* at Planning for Human Settlements in Disaster-prone Areas Seminar: United Nations Centre for Human Settlements (Habitat), Nairobi.

Lean, Geoffrey (1988): Environment Report *The Observer* 9 October., London.

Lean, Geoffrey (1994): Early Warnings from Small Islands *Choices* UNDP. New York.

Lewis, James (1977): Some aspects of disaster research *Disasters* 1/3; pp 241–244.

Lewis, James (1978): *Mitigation and preparedness for natural disasters in the Kingdom of Tonga* Report to the Ministry of Overseas Development (ODA).

Lewis, James (1979a): The Vulnerable State: An alternative view In *Disaster Assistance: Appraisal, reform and new approaches* (Stephens, L H and Green, S J: Eds) Chapter 5 New York University Press.

Lewis, James (1979b): Volcano in Tonga *Journal of Administration Overseas* XVIII/2 ODM/HMSO. London.

Lewis, James (1980): *Analysis of Olsen (1977)*: Personal research files. Manuscript.

Lewis, James (1981a): *Assistance préparatoire pour la reconstruction des établissements humains dans la région d'El Asnam* Rapport de Mission (avec F D Mokrane). République Algérienne Démocratique et Populaire: Commission Interministérielle pour l'Etude de la Reconstruction d'El Asnam Centre des Nations Unies pour les Etablissements Humains (CNUEH: Habitat), Nairobi.

Lewis, James (1981b): Some perspectives on natural disaster vulnerability in Tonga *Pacific Viewpoint* 22/2 pp 145–162. Victoria University of Wellington.

Lewis, James (1981c): The Sri Lanka Cyclone 1978; Socio-economic analysis of housing destruction *Marga* 6/2. Colombo. 1981 (Also published as Cyclone destruction in Sri Lanka; Some socio-economic analysis *Disaster Management* 2/3. New Delhi).

Lewis, James (1982a): *A Three-part Proposal: Three stages for natural disaster mitigation strategy: Integrated with socio-economic development at national, provincial and local levels* Mimeo. Datum International. Marshfield.

Lewis, James (1982b): Natural Disaster Mitigation: Environmental approaches in Tonga and Algeria *The Environmentalist* 2; pp 233–246.

Lewis, James (1982c): *The Social and Economic Effects of Natural Disasters on Least-developed Island Countries: With special reference to Antigua; Cape Verde Islands; Comoro Islands; Maldive Islands; and Western Samoa* Report to UNCTAD VI 1983. UNDRO/UNCTAD.

Lewis, James (1983a): *Possible Imminent Severe Water and Food Shortage; Ha'apai Islands, Tonga* A Report to the United Nations Development Programme Resident Representative, Suva, Fiji; 13 June.

Lewis, James (1983b): The Long-term Implications of Hurricane Isaac (March 1982) Mission Report; United Nations Centre for Human Settlements (Habitat), Nairobi.

Lewis, James (1983c): Change, and vulnerability to natural hazard: Chiswell, Dorset *The Environmentalist* 3 pp277–287.

Lewis, James (1984a): Multi-hazard history in Antigua *Disasters* 8/3.

Lewis, James (1984b): Vulnerability to a cyclone: Damage distribution in Sri Lanka *Ekistics: The problems and science of human settlements* 51/308. Athens.

Lewis, James (1984c): Environmental interpretations of natural disaster mitigation: The crucial need Guest Editorial *The Environmentalist* 4 pp 177–180.

Lewis, James (1985): *Project identification in Hazardous Environments* Mimeo. Datum International. August.

Lewis, James (1986): Letter; *The Guardian* (radiation risk) 12 May.

Lewis, James (1987a): *Vulnerability and Development – and the Development of vulnerability: A case for management* Paper presented at the *Development Studies Association Annual Conference*, University of Manchester.

Lewis, James (1987b): *Development in Earthquake Areas and the Management of vulnerability* Paper delivered at the Annual Colloquium of Research in Progress Housing and Physical Development in Algeria Centre for Architectural Research and Development Overseas, University of Newcastle.

Lewis, James (1987c): Risk, vulnerability and survival; Implications for people, planning and civil defence *Local Government Studies* August pp 75–93.

Lewis, James (1988a): An open letter in response to Confronting Natural Disasters; An International Decade for Natural Hazard Reduction *Natural Hazards Observer* XII/4 March.

Lewis, James (1988b): *Sea Level Rise: Tonga; Tuvalu; (Kiribati): Report of a rapid field reconnaissance mission*. Commonwealth Expert Group on Climate Change and Sea-level Rise, Commonwealth Secretariat.

Lewis, James (Ed: 1988c): *Settlement Reconstruction Post-war* Report of a Two Day Workshop. Institute for Advanced Architectural Studies, University of York.

Lewis, James (1989a): Sea-Level Rise: Some implications for Tuvalu *Ambio* 18:8.

Lewis, James (1989b): *Affordability and Participation, Need and Vulnerability: Post-cyclone rehousing in Tonga* Sixth Inter-Schools Conference on Development Centre for Development Planning Studies, University of Sheffield.

Lewis, James (1990): Small States Conference on Sea Level Rise *The Environmentalist* 10:2.

Lewis, James (1991a): *The Relief Syndrome: Post World War II and sequentialism* Mineo. Datum International.

Lewis, James (1991b): Tropical cyclones and island states In *Shelter, Settlements, Policy and the Poor* Intermediate Technology Publications.

Lewis, James (1994a): *Disasters, Conflict and Recovery: What kind of development for sustainability and vulnerability reduction?* Paper prepared for People, Place and Development International Symposium, Centre for Architectural Research and Development, University of Newcastle.

Lewis, James (1994b): *Vulnerability Reduction, Survival and Sustainability: What kind of development?* Paper presented at the *Seminar on Civil Strife and Relief within the Context of the Continuum from Relief to Development* The Institute of Social Studies, The Hague. July.

Lewis, James (1994c): Volcanoes and war in Papua New Guinea *Stop Disasters* 21 (September/October) IDNDR Secretariat.

Lewis, James (1994d): Earthquakes and war in Algeria *Stop Disasters* 21 (September/October) IDNDR Secretariat.

Lewis, James (1994e): The vulnerability of small island states to sea level rise; The need for holistic strategies *Disasters* 14:3.

Lewis, James (1996): *Sea Level Rise; A case of cost transfer and responsibility* Msc Module in Environmental Science, Policy and Planning: Management of the Commons. International Centre for the Environment/Centre for Development Studies, University of Bath.

Lewis, James (1997): Development, vulnerability and disaster reduction; Bangladesh cyclone shelter projects and their implications In *Reconstruction After Disaster: Issues and practices* (Awotona, Adenrele: Ed). Chapter 4 Ashgate).

Lewis, James (1998): Volcanic eruption on Monserrat: Perspectives on risk-management and vulnerability. Submitted for publication.

Lewis, James; Carter, Tom (1986): *Beyond Conventional Disaster Management* Mimeo. Datum International. Marshfield.

Macrae, Joanna; Zwi, Anthony B (1992): Food as an Instrument of War in Contemporary African Famines: A Review of the Evidence *Disasters* 14/4.

Madrid (1995): *The Madrid Declaration:* Declaration of the Humanitarian Summit Meeting held in Madrid; signed by the Administrator USAID; Executive Directors of UNICEF & WFP; European commissioner for Humanitarian Aid; United Nations Under-Secretary General responsible for Humanitarian Affairs; United Nations High Commissioner for Refugees; and the Presidents of the Liaison Committee of Development NGOs to the European Union, Medcins sans Frontiers, International Committee of the Red Cross, and Interaction (American Council for Voluntary International Action).

Maskrey, Andrew (1989): Disaster Mitigation; A community based approach *Development Guidelines No 3*. Oxfam. Oxford.

Maslin, Mark (1998): Global Warming: Wave after wave *The Guardian* 7 October.

McEntire, David A (1998): Balancing international approaches to disaster: rethinking prevention instead of relief *Australian Journal of Emergency Management* 13/2 Winter. Macedon.

Mensching, H G; Geist, H; Kishk, M A (1997): Ecological marginality: Natural and human factors of raising vulnerabilities to disasters *Stop Disasters* 31/1. IDNDR. Geneva.

Mitchell, James K; Devine, Neal; Jagger, Kathleen (1989): A contextual model of natural hazard *Geographical Review* 79/4 October.

Morse, Bradford (1977): *Practice, Norms and Reform of International Humanitarian Rescue Operations* Lectures at the Hague Academy of International Law (unofficial version).

OAS (1990): *Primer on Natural Hazard Management in Integrated Regional Development Planning* Department of Regional Development and Environment, Executive Secretariat for Economic and Social Affairs of the Organisation of American States. Washington DC.

167

OECD (1997): *DAC Guidelines on Conflict, Peace and Development Co-operation* Development Assistance Committee. Paris.

ODI (1998:1): *The State of the International Humanitarian System* Briefing Paper. Overseas Development Institute. London.

ODI (1998:2): *The UK White Paper on International Development – and Beyond* Briefing Paper. Overseas Development Institute. London.

OFDA (1978): Sri Lanka – Cyclone. Situation Report 1: 27 November; 2: 25 November; 3: 12 December. Washington DC., Department of State.

OFDA (1997a): *Linking Relief to Development: Disaster response with foresight* Annual Report Office of US Foreign Disaster Assistance, Bureau for Humanitarian Response, US Agency for International Development. Washington DC.

OFDA (1997b): *Linking Development to Relief: The integral role of prevention, mitigation, preparedness and planning* Annual Report Office of US Foreign Disaster Assistance, Bureau for Humanitarian Response, US Agency for International Development. Washington DC.

ONRS (1980): *Dossiers Documentaires No 14* Centre de Recherches en Architecture et Urbanisme, Organisation National de la Recherche Scientifique, Algiers.

Olsen, Robert A; Olsen, Richard Stuart (1977): The Guatemala Earthquake of 4 February 1976: Social science observations and research suggestions *Mass Emergencies* 2 pp 69–81.

Park, Wayne; Bender, Stephen (1990): *The Vulnerability of the Electrical Sector to Natural Hazards in Costa Rica* The World Bank/Organisation of the American States. June.

Parker, Dennis; Islam, Nabiul; Ngai, Weng Chan (1997): Reducing vulnerability following flood disasters: Issues and practices Chapter 3 in *Reconstruction After Disaster: Issues and practices* (Awotona, Adenrele: Ed) Ashgate.

Pelanda, Carlo (1981): *Disaster and sociosystemic vulnerability* The social and economic aspects of earthquakes and planning to mitigate their effects Third International Conference. Bled, Yugoslavia.

Pitman (1997): Population and vulnerability *Project communication* World Bank. Washington DC.

Quarantelli, Enrico L (1985): An assessment of conflicting views on mental health: the consequences of traumatic events In Figley, C (ed) *Trauma and its wake* Brunner/Mazel. New York. pp 182–220.

Quaranteli, Enrico L (1986): Planning and management for the prevention and mitigation of natural disasters, especially in a metropolitan context: Initial questions and issues which need to be addressed *Planning for Crisis Relief* International Seminar; United Nations Centre for Regional Development, Nagoya.

Rahman, Atiur (1991): Development responses to natural disaster. Chapter 11 in *Participatory Development: Learning from South Asia* (Eds: Wignaraja, Ponna; Hussain, Akmal; Sethi, Harsh; Wignaraja, Ganeshan) United Nations University Press/Oxford University Press. Tokyo/Karachi.

Rawson, C B (1868): *Report on the Bahamas Hurricane of October 1866, with a Description of the City of Nassau* NP (Office of the Nassau Guardian, 1868. Nassau Public Library).

Rio (1992): *Agenda 21 Final Documents*; Earth Summit. Rio de Janeiro. June.

ReliefWeb/Agence France-Presse (1998): *Hurricane Mitch in Central America.* 2 November.

Skultans, Vieda (1979): *English Madness; Ideas on insanity 1580–1890* Routledge & Kegan Paul. London.

Smith, Brian (1982): *Centralisation and Underdevelopment* Seminar notes: Centre for Development Studies, University of Bath.

Snarr, D Neil; Brown, E Leonard (1979): Permanent post-disaster housing in Honduras: aspects of vulnerability to future disasters *Disasters* 3/3 pp 287–292

Stevenson, Robert Louis (1892): The Hurricane; March 1889. Chapter X in *A Footnote to History; Eight years of trouble in Samoa* Cassell.

Stewart (1993): War and underdevelopment: can economic analysis help reduce the costs? Journal of International Development 3/4 pp357–380. John Wiley.

Sutton, Keith (1969): Algeria's population growth 1954–1966 *Geography* 54/244 (July) pp 332–325.

Sutton, Keith (1978): The progress of Algeria's Agrarian Reform and its settlement implications *Maghreb Review* 3/5–6 pp 10–16.

Sutton, K; Lawless, Richard I (1978): Population regrouping in Algeria; traumatic change and the rural settlement pattern *Transactions of the Institute of British Geographics* 3/3 pp 331–350.

Tapner, Vic (1977): Karachi: City in chaos *Building Design* 18 November.

Thaman, R R; Meleisea, M; Makasiale, J (1979): Agricultural diversity and traditional knowledge as insurance against natural disasters. Summary record of the Natural Disaster Prevention, Preparedness and Rehabilitation Meeting. Annex IO. South Pacific Bureau for Economic Co-operation (South Pacific Forum). Suva.

Turton, David (1992): Warfare, Vulnerability and Survival: A case from Southwestern Ethiopia *Disasters* 15:3.

UNCED (1993): *Report of the United Nations Conference on Environment and Development* Rio de Janeiro 1992. United Nations.

UNDRO (1987): *List of National Officials Responsible for the Management of Disasters and Other Emergencies, emergency plans and disaster legislation* Office of the United Nations Disaster Relief Co-ordinator, Geneva.

UNEP (1988): *Report of the Joint Meeting on Implications of Climate Change* Split. October.

UNESCO (1977): *Population, Resources and Development in the Eastern Islands of Fiji: Information for decision makers* General Report No 1 and Ecology and Rational Use of Land Systems Project No 7 Man and the Biosphere Programme. Paris.

UNRISD (1993): *Rebuilding Wartorn Societies* Report of the workshops on The Challenge of Rebuilding Wartorn Societies and The Social Consequences of the Peace Process in Cambodia. United Nations Research Institute for Social Development. Geneva.

WCED (1987): *Our Common Future* The World Commission on Environment and Development. Oxford University Press.

Wells, Sue; Edwards, Alasdair (1989): Gone with the waves *New Scientist* November.

Westgate, K N (1975): A disaster history of Tonga: 1909–1963 Mimeo. Disaster Research Unit, University of Bradford.

Westgate, K N; O'Keefe, P 0 (1976): *The Human and Social Implications of Earthquake Risk for Developing Countries: Towards an integrated mitigation strategy* Intergovernmental Conference on the Assessment and Mitigation of Earthquake Risk United Nations Educational, Scientific and Cultural Organisation (UNESCO), Paris.

Westgate, Westgate, Ken (1996): Disaster management as a development activity: The approach of the UN Disaster Management Training Programme *DHA News* April/May.

Westing, Arthur H (1990): *Environmental Hazards of War: Releasing dangerous forces in an industrialised world.* Sage. London.

Whitehouse, Christopher (1996): *Human Rights and Basic Needs in Development: Are they universal? Are they universalisable?* Occasional Paper 06/96. Centre for Development Studies, University of Bath. December.

Winchester, Peter (1986): *Vulnerability and Recovery in Hazard Prone Areas* Paper presented to the Middle East and Regional Conference on Earthen and Low Strength Masonry Buildings in Seismic Areas: Middle East Technical University, Ankara, Turkey.

Winchester, Peter (1992): *Power, Choice and Vulnerability* James & James.

Wintzel, A. K (1979): The vulnerable society *Alternative Patterns of Development and Lifestyles* UNEP/ECE Regional Seminar. Ljubljana. (Also as *The Vulnerable Society* The Secretariat for Future Studies, Stockholm. c1981).

Wolde Miriam, Mesfin (1986): *Rural Vulnerability to Famine in Ethiopia* Intermediate Technology Publications. London.

World Bank (1980): *1980 World Bank Atlas* World Bank, Washington DC.

World Bank (1996): Towards social sustainability *Environment Matters Annual Review*; July 1995–June 1996 (FY 96). Fall.

Zulueta, Felicity de (1993): *From Pain to Violence: The traumatic roots of destructiveness* Whurr. London.

Index

drought mortality, 81

earthquake, 3, 4, 12 - 13, 15, 17, 18, 25, 30, 31, 33, 36, 38, 40, 45, 50, 51, 52, 54, 56, 60, 62, 63, 65, 67, 68, 69, 74, 75, 76, 77, 78, 82, 84, 85, 88, 125, 140, 151, 159
earthquake destruction, 76 -77
earthquake loan, 80, 85
earth sciences, 109
East Africa, xiv
East Caribbean Common Market, 84
education, 7
Egypt, 28
emancipation (slavery), 76, 77
embankments, 16, 26, 41
emergency shelter, 63, 64
energy, 156
engineering, 152
England, 26, 48, 50, 52, 82, 109
  Abbotsbury/Swannery, 109, 114
  Central Government, 117
  Chesil Beach, 109, 111, 112, 113, 114, 121
  Coastal Protection Act 1949, 120
  Dorking, 121
  Dorset, 109, 121, 131
  Fleet, 112
  the Fleet, 109, 114, 115
  Housing Improvement Act, 119
  Institute of Oceanographic Sciences, 122
  Land Drainage Act 1976, 120
  Langton Herring, 111
  London, 51
  Lyme Bay, 109
  Portland/Isle of Portland, 52, 109, 110, 111, 112, 113, 115, 121, 122
    Chesil Cove, 116
    Chiswell, 109 - 121, 131
    Chiswell Residents' Action Group (CRAG), 117, 118, 119, 122
    Cove Inn, 114
    Fortuneswell, 109, 112
    Naval Helicopter Station, 115
    Portland Mere, 115
    Victoria Square, 114, 115, 116
    West Portland Borough Council, 118
  Portland Harbour, 112, 115, 121
  Queen's Bench Division, 121
  Wessex Water Authority, 116, 117, 118, 122
  West Bay, 121
  Weymouth, 109, 113
  Weymouth Bay, 116
  Weymouth & Portland Borough Council, 116, 121
    Policy Resources Committee, 117
  Wyke Regis, 111
English Channel, 112
environmental conservation, 120, 141
environmental damage, 26
environmental health, 22, 36, 66
environmental impact assessment (EIA), 136
environmental management, 135 - 136
environmentalists, 152
epidemic, 74
equilibrium profile, 111

equitable development/equitability/inequitability, xviii, 40, 137 -138, 148, 149, 151, 157
erosion, 28, 158
Ethiopia, 34, 39
  Bodi, 34
  Mursi, 34
European Bank for Reconstruction & Development, 131
European Economic Commission Disaster Fund, 118
European Union, 132
evacuation, 54 - 59, 63
evaluation, 150, 153
explosion, 3, 40
famine, 125, 161
feasting, 71 -72

Fiji, 23, 67 (see also: Hurricane Bebe)
  Na'adi, 57
financial participation, 64, 159
fire, 3, 30, 51, 52, 77, 84, 85, 125, 134
fisheries/fishing/fishermen, 19, 69, 84, 100, 112
flash flooding, 31
floods/flooding, 5, 13, 15, 21, 29, 45, 48, 51, 52, 88, 111, 114, 115, 116, 117, 118, 122, 125, 134, 151, 152, 158, 159
flood-tide, 114
flood warnings, 114 (see also: warnings)
flood water, 114
food/food supplies, 7, 27, 36, 40, 64, 69, 83, 155
food crops, 28, 31, 40, 51, 67
food shortage, 71
France, 32, 56
  Paris, 37, 85
French Colonial Government, 37
French colonists/colons, 27, 32, 85
French Metropolitan Government, 37
French West Indian Colonies, 82

geography/geographers, xiv, 161
Germany, 34
German New Guinea, 34
Grameen Bank, 141
Great Britain, 76
  Department of the Colonies, 85
  HM Government, 79, 83
  HM Treasury, 79
  Parliament, 78, 82
Greece, 88
Guatemala, 4
  earthquake, 18, 60
Gulf of Mannar, 87

Haiti, 27
health services, 7, 39, 139, 140, 154, 155
  maternity & child care, 154
  preventive medicine, 154
Hispaniola, 28
Holland, 56
homelessness, 3, 13, 16, 27, 45, 46, 48, 151
Honduras, 19, 45 (see also: Hurricane Mitch)
  Islas de la Bahia, 19
housing/rehousing/housing reconstruction, 16, 137, 152
housing construction, 21, 30
housing damage/destruction, 3, 15, 16, 31, 62,

173

Cabinet, 64
Central Planning Unit, 72
Disaster Relief Committee, 72
'Eua, 55, 56, 58, 66
Angaha, 56
folklaw, 63
Fonua Fo'ou, 62
Government of Tonga, 60, 63
Ha'apai Group, 31, 50, 60, 63, 54, 66, 72
Hurricane Relief Committee, 64, 65, 68, 72
Jack-in-a-Box Island, 62
Kao, 20
Kavaliku, The Hon Langi, 64
Malekamu, John, 56
Ministry of Finance, 72
Ministry of Health, 70
Ministry of Works, 64
Nuku'alofa, 54, 57, 58, 60, 66
Niua Fo'ou, 20, 38, 39, 54–59, 63, 66
    Angaha, 56
    Futu, 55
    Petani, 55
Niua Group, 60
Niuatopatapu, 54, 62
Population, 65–66
Prime Minister, 64, 72
Queen Salote, 55
Royal Feast, 72
Takai, Moeake, 54, 56, 58
Third Development Plan, 72
Tin Can Mail Island, 54
Tongatapu, 54, 57, 58, 60, 65
Tui Tonga, 71
Vava'u Group, 54, 58, 60, 63
Ve'ehala, The Hon, 63
Tonga Deep, 65
Tortula, 82
traditional building construction, 67
traditional responses, 63
transport/communications, 49, 64, 107, 146, 150, 154, 156
Treaty of Berlin, 25, 75
tropical cyclones/hurricanes/typhoons, 3, 7, 13, 15, 22, 26, 28, 29, 30, 31, 36, 38, 40, 45, 48, 50, 52, 54, 60, 62, 63, 64, 65, 66, 67, 69, 71, 74, 80, 81, 82, 83, 85, 87–107, 125, 139, 141, 151, 159 *(see also hurricanes by name)*
Tropical Cyclone No 21, 87
tsunami, 3, 54, 62, 63
Tunisia, 32
Tuvalu (Ellice Islands), 20–21, 49, 62
    Funafuti, 21, 22, 26, 49, 62
    Trust Fund, 20
    Vaiaku, 22

Uganda, 39
United Nations, 131
United Nations Conference on Trade & Development (UNCTAD), 74
United Nations Development Programme (UNDP), 131, 135
United Nations Office of the Disaster Relief Co-ordinator (UNDRO)/Department of Humanitarian Affairs (DHA), 74, 131

United Nations Office for the Co-ordination of Humanitarian Affairs (OCHA), 131
United Nations Relief & Rehabilitation Administration (UNRRA), 131
United Nations Research Institute for Social Development (UNRISD), 94
United Nations Secretary General, 131
United Kingdom, 20
United States of America/USA, xii, 4, 78, 82, 132, 161
    Texas, 45, 56
    US Government, 57
    US Navy, 57
    US Pacific Fleet, 25
University of Aukland, 63
University of Bath, Institute for International Policy Analysis/Centre for Development Studies, x, 7
University of Bradford/Disaster Research Unit, x, 7
University of Colorado/Natural Hazards Group, x, xiv, 122
University of Toronto, x, xiv
urban impact/damage, 17, 49, 90, 106, 107, 158
urban vulnerability, 147

Vietnam, 45, 46
Virgin Islands, 82
    St Croix, 4
vitamin tablets, 70
vocational vulnerability, 16
volcano/volcanic eruption, 3, 33, 34, 49, 52–54, 56, 62, 63, 75, 125, 152
vulnerability analysis/assessment (VA), 131, 149, 159
vulnerability indicators, 7, 14 -15, 90, 94–97, 102–105
vulnerability matrix, 14

war/warfare/war damage/enemy action, 34–35, 36, 41, 75, 113, 114, 120, 131, 132, 137, 138, 145, 148, 149, 151, 158, 163 *(see also: conflict)*
warnings, 31, 41, 83, 127, 146, 152 *(see also: flood warnings)*
water supply, 22, 31, 32, 36, 38, 40, 51, 54, 64, 66, 84, 155–157 *(see also: drinking water)*
Western Samoa, 25, 54, 65, 74, 75
    Apia, 25, 65, 75
West Indian Colonies, 82
West Indian & Panama Telegraph Company, 83
women in development, 152, 154
World Bank, 131
World Commission on Environment & Development (WCED), 135
world wars, 63
World War I, 34
World War II, 21, 25, 34, 36, 131

Yugoslavia, xvii, 88
Y2K, 25

Zanzibar, 30